Historical Perspectives from a Nation Divided:
Accounts of the Battle of South Mountain

BLUE MUSTANG
P R E S S

Blue Mustang Press
Boston, Massachusetts

ISBN 978-1-935199-13-7
PUBLISHED BY BLUE MUSTANG PRESS
www.BlueMustangPress.com
Boston, Massachusetts

Printed in the United States of America

Historical Perspectives from a Nation Divided:
Accounts of the Battle of South Mountain

Reprinting selections from:

MILITARY REMINISCENCES
OF THE CIVIL WAR
by
JACOB DOLSON COX, A.M., LL.D.
Formerly Major-General commanding Twenty-Third Army Corps
VOLUME I
Published in 1900

~~~~~

THE LIFE, PUBLIC SERVICES AND SELECT
SPEECHES OF RUTHERFORD B. HAYES
by
J. Q. HOWARD
Published in 1876

~~~~~

THREE YEARS IN THE SIXTH CORPS.
A CONCISE NARRATIVE OF EVENTS IN THE ARMY
OF THE POTOMAC, FROM 1861 TO THE CLOSE OF
THE REBELLION, APRIL, 1865.
by
GEORGE T. STEVENS,
Surgeon of the 77th Regiment New York Volunteers
Published in 1866

~~~~~

A LIFE OF GEN. ROBERT E. LEE.
by
JOHN ESTEN COOKE
Published in 1876

~~~~~

HISTORY OF KERSHAW'S BRIGADE
by
D. AUGUSTUS DICKERT
Published in 1907

Introduction to this Series of Books

This series of books, "Historical Perspectives from a Nation Divided" seeks to offer different - often contemporary - views of related historical events. While other volumes in this series divide the accounts between North and South, this entry into the series seeks to gather various accounts, by both those who were there at South Mountain and those who assembled the facts later.

There is a great wealth of histories, recollections, regimental diaries, and other such types of published works relating to the people and operations of the American Civil War that were written by the then-living doers of those deeds. The surge in such writings, starting immediately after the War's end and peaking in the early years of the 20th century (when these writers were facing their own mortality) has left us with countless recountings of similar events. This series hopes to bring many of those texts back to those who find interest in them, both on an entertainment level and on a research level.

The "cast of characters" in this volume includes many of the "big names" of the war, and even a future president. Yet these accounts and narratives brings these personages down to real life level and away from the idealized accounts in history books.

These beliefs are well represented in the works published here and it's an unquestioned value that these treasures still exist to provide a window into the construction of history.

Table of Contents

Also included are photographs, maps, and other illustrations related to events surrounding the Battle of South Mountain.

Robert E. Lee

Special Orders, No. 191

Much as been made of the careless losing and lucky finding of a copy of Lee's plans for troop movements wrapped around three cigars in a field in Frederick County. Maj. Gen. George B. McClellen thought them of great import, even if he did not utilize them to their fullest fearing a trap. Others have postulated that none of the battles-neither South Mountain nor Antietam-would have ended any differently had the orders never been lost (or found).

Naturally, these questions are impossible to answer. What is known is that on the morning of September 13, 1862, Corporal Barton W. Mitchell of the 27th Indiana Volunteers, of the Union XII Corps, chanced upon an envelope that contained three cigars wrapped with a leaf of paper...Special Orders No. 191...in a field that General D. H. Hill's division had recently vacated.

Realizing the potential of this document (another lucky happenstance), the corporal submitted it to his superior and it went up the chain of command until it was given to McClellan himself. McClellen felt the information was such that it would allow him to beat Lee soundly.

The orders read:

Special Orders, No. 191
Hdqrs. Army of Northern Virginia
September 9, 1862

1. The citizens of Fredericktown being unwilling while overrun by members of this army, to open their stores, to give them confidence, and to secure to officers and men purchasing supplies for benefit of this command, all officers and men of this army are strictly prohibited from visiting Fredericktown except on business, in which cases they will bear evidence of this in

writing from division commanders. The provost-marshal in Fredericktown will see that his guard rigidly enforces this order.

2. Major Taylor will proceed to Leesburg, Virginia, and arrange for transportation of the sick and those unable to walk to Winchester, securing the transportation of the country for this purpose. The route between this and Culpepper Court-House east of the mountains being unsafe, will no longer be traveled. Those on the way to this army already across the river will move up promptly; all others will proceed to Winchester collectively and under command of officers, at which point, being the general depot of this army, its movements will be known and instructions given by commanding officer regulating further movements.

3. The army will resume its march tomorrow, taking the Hagerstown road. General Jackson's command will form the advance, and, after passing Middletown, with such portion as he may select, take the route toward Sharpsburg, cross the Potomac at the most convenient point, and by Friday morning take possession of the Baltimore and Ohio Railroad, capture such of them as may be at Martinsburg, and intercept such as may attempt to escape from Harpers Ferry.

4. General Longstreet's command will pursue the same road as far as Boonsborough, where it will halt, with reserve, supply, and baggage trains of the army.

5. General McLaws, with his own division and that of General R. H. Anderson, will follow General Longstreet. On reaching Middletown will take the route to Harpers Ferry, and by Friday morning possess himself of the Maryland Heights and endeavor to capture the enemy at Harpers Ferry and vicinity.

6. General Walker, with his division, after accomplishing the object in which he is now engaged, will cross the Potomac at Cheek's Ford, ascend its right bank to Lovettsville, take possession of Loudoun Heights, if practicable, by Friday morning, Key's Ford on his left, and the road between the end

of the mountain and the Potomac on his right. He will, as far as practicable, cooperate with General McLaws and Jackson, and intercept retreat of the enemy.

7. General D. H. Hill's division will form the rear guard of the army, pursuing the road taken by the main body. The reserve artillery, ordnance, and supply trains, &c., will precede General Hill.

8. General Stuart will detach a squadron of cavalry to accompany the commands of Generals Longstreet, Jackson, and McLaws, and, with the main body of the cavalry, will cover the route of the army, bringing up all stragglers that may have been left behind.

9. The commands of Generals Jackson, McLaws, and Walker, after accomplishing the objects for which they have been detached, will join the main body of the army at Boonsborough or Hagerstown.

10. Each regiment on the march will habitually carry its axes in the regimental ordnance—wagons, for use of the men at their encampments, to procure wood &c.

By command of General R. E. Lee
R.H. Chilton, Assistant Adjutant General

The following account is taken from:

MILITARY REMINISCENCES OF THE CIVIL WAR

by

JACOB DOLSON COX, A.M., LL.D.
Formerly Major-General commanding Twenty-Third Army Corps

VOLUME I.

APRIL 1861—NOVEMBER 1863

JACOB D. COX. MAJ. GEN.

ÆT 34

PREFACE

My aim in this book has been to reproduce my own experience in our Civil War in such a way as to help the reader understand just how the duties and the problems of that great conflict presented themselves successively to one man who had an active part in it from the beginning to the end. In my military service I was so conscious of the benefit it was to me to get the personal view of men who had served in our own or other wars, as distinguished from the general or formal history, that I formed the purpose, soon after peace was restored, to write such a narrative of my own army life. My relations to many prominent officers and civilians were such as to give opportunities for intimate knowledge of their personal qualities as well as their public conduct. It has seemed to me that it might be useful to share with others what I thus learned, and to throw what light I could upon the events and the men of that time.

As I have written historical accounts of some campaigns separately, it may be proper to say that I have in this book avoided repetition, and have tried to make the personal narrative supplement and lend new interest to the more formal story. Some of the earlier chapters appeared in an abridged form in "Battles and Leaders of the Civil War," and the closing chapter was read before the Ohio Commandery of the Loyal Legion. By arrangements courteously made by the Century Company and the Commandery, these chapters, partly re-written, are here found in their proper connection.

Though my private memoranda are full enough to give me reasonable confidence in the accuracy of these reminiscences, I have made it a duty to test my memory by constant reference to the original contemporaneous material so abundantly preserved in the government publication of the Official Records of the Union and Confederate Armies. Where the series of these records is not given, my references are to the First Series, with the abbreviation O. R., and I have preferred to adhere to the official designation of the volumes in parts, as each volume then includes the documents of a single campaign.

J. D. C.

NOTE.—The manuscript of this work had been completed by General Cox, and placed in the hands of the publishers several weeks before his untimely death at Magnolia, Mass., August 4, 1900. He himself had read and revised some four hundred pages of the press-work. The work of reading and revising the remaining proofs and of preparing a general index for the work was undertaken by the undersigned from a deep sense of obligation to and loving regard for the author, which could not find a more fitting expression at this time. No material changes have been made in text or notes. Citations have been looked up and references verified with care, yet errors may have crept in, which his well-known accuracy would have excluded. For all such and for the imperfections of the index, the undersigned must accept responsibility, and beg the indulgence of the reader, who will find in the text itself enough of interest and profit to excuse many shortcomings.

WILLIAM C. COCHRAN.
CINCINNATI, October 1, 1900.

CHAPTER XIII
SOUTH MOUNTAIN

March through Washington—Reporting to Burnside—The Ninth Corps—Burnside's personal qualities—To Leesboro—Straggling—Lee's army at Frederick—Our deliberate advance—Reno at New Market—The march past—Reno and Hayes—Camp gossip—Occupation of Frederick—Affair with Hampton's cavalry—Crossing Catoctin Mountain—The valley and South Mountain—Lee's order found—Division of his army—Jackson at Harper's Ferry—Supporting Pleasonton's reconnoissance—Meeting Colonel Moor—An involuntary warning—Kanawha Division's advance—Opening of the battle—Carrying the mountain crest—The morning fight—Lull at noon—Arrival of supports—Battle renewed—Final success—Death of Reno—Hooker's battle on the right—His report—Burnside's comments—Franklin's engagement at Crampton's Gap.

Late in the night of the 5th I received orders from McClellan's headquarters to march from my position on Upton's Hill through Washington toward Leesboro, [*Note: Leesboro, a village of Maryland eight or ten miles north of Washington, must be distinguished from Leesburg in Virginia.*] as soon as my pickets could be relieved by troops of McDowell's corps. [*Note: Official Records, vol. xix. pt. ii. p. 183; vol. li. pt. i. p. 789.*] My route was designated as by the road which was a continuation northward of Seventh Street, and I was directed to report to General Ambrose E. Burnside, commanding right wing, whose headquarters were in the suburbs of the city on that road. This was in accordance with my wish, expressed to McClellan that I might have active field work. For two or three days we were not attached to a corps, but as the organization of the army became settled we were temporarily assigned to the Ninth, which had been Burnside's, and had been with him in North Carolina. During this campaign it was commanded by Major-General Jesse L. Reno, who had long had a division in it, and had led the corps in the recent battle. We marched from Upton's Hill at

19

daybreak of the 6th, taking the road to Georgetown by Ball's Cross-Roads. In Georgetown we turned eastward through Washington to Seventh Street, and thence northward to the Leesboro road. As we passed General Burnside's quarters, I sent a staff officer to report our progress. It was about ten o'clock, and Burnside had gone to the White House to meet the President and cabinet by invitation. His chief of staff, General J. G. Parke, sent a polite note, saying we had not been expected so soon, and directed us to halt and bivouac for the present in some fields by the roadside, near where the Howard University now is. In the afternoon I met Burnside for the first time, and was warmly attracted by him, as everybody was. He was pre-eminently a manly man, as I expressed it in writing home. His large, fine eyes, his winning smile and cordial manners, bespoke a frank, sincere, and honorable character, and these indications were never belied by more intimate acquaintance. The friendship then begun lasted as long as he lived. I learned to understand the limitations of his powers and the points in which he fell short of being a great commander; but as I knew him better I estimated more and more highly his sincerity and truthfulness, his unselfish generosity, and his devoted patriotism. In everything which makes up an honorable and lovable personal character he had no superior. I shall have occasion to speak frequently of his peculiarities and his special traits, but shall never have need to say a word in derogation of the solid virtues I have attributed to him. His chief-of-staff, General Parke, was an officer of the Engineers, and one of the best instructed of that corps. He had served with distinction under Burnside in North Carolina, in command of a brigade and division. I always thought that he preferred staff duty, especially with Burnside, whose confidence in him was complete, and who would leave to him almost untrammelled control of the administrative work of the command.

On September 7th I was ordered to take the advance of the Ninth Corps in the march to Leesboro, following Hooker's corps. It was my first march with troops of this army, and I was shocked at the straggling I witnessed. The "roadside brigade," as we called it, was often as numerous, by careful estimate, as our own column moving in the middle of the road. I could say of the men of the Kanawha division, as Richard Taylor said of his Louisiana brigade with Stonewall Jackson, that they had not yet learned to straggle. [*Note: See Taylor's "Destruction and Reconstruction," p. 50, for a curious interview with Jackson.*] I tried to prevent their learning it. We had a roll-call immediately upon halting after the march, and another half an hour later, with prompt reports of the result. I also assigned a field officer and medical officer to duty at the rear of the column, with ambulances for those who became ill and with punishments for the rest. The result was that, in spite of the example of others, the division had no stragglers, the first roll-call rarely showing more than twenty or thirty not answering to their names, and the second often proving every man to be present. [*Note: See letters of General R. B. Hayes and General George Crook, Appendix B.*] In both the Army of the Potomac and the Army of Northern Virginia the evil had become a most serious one. After the battle of Antietam, for the express purpose of remedying it, McClellan appointed General Patrick Provost-Marshal with a strong provost-guard, giving him very extended powers, and permitting nobody, of whatever rank, to interfere with him. Patrick was a man of vigor, of conscience, and of system, and though he was greatly desirous of keeping a field command, proved so useful, indeed so necessary a part of the organization, that he was retained in it against his wishes, to the end of the war, each commander of the Army of the Potomac in turn finding that he was indispensable. [*Note: I have discussed this subject also in a review of Henderson's Stonewall Jackson, "The Nation," Nov. 24, 1898, p. 396.*]

The Confederate army suffered from straggling quite as much, perhaps, as ours, but in a somewhat different way. At the close of the Antietam campaign General Lee made bitter complaints in regard to it, and asked the Confederate government for legislation which would authorize him to apply the severest punishments. As the Confederate stragglers were generally in the midst of friends, where they could sleep under shelter and get food of better quality than the army ration, this grew to be the regular mode of life with many even of those who would join their comrades in an engagement. They were not reported in the return of "effectives" made by their officers, but that they often made part of the killed, wounded, and captured I have little doubt. In this way a rational explanation may be found of the larger discrepancies between the Confederate reports of casualties and ours of their dead buried and prisoners taken.

The weather during this brief campaign was as lovely as possible, and the contrast between the rich farming country in which we now were, and the forest-covered mountains of West Virginia to which we had been accustomed, was very striking. An evening march, under a brilliant moon, over a park-like landscape with alternations of groves and meadows which could not have been more beautifully composed by a master artist, remains in my memory as a page out of a lovely romance. On the day that we marched to Leesboro, Lee's army was concentrated near Frederick, behind the Monocacy River, having begun the crossing of the Potomac on the 4th. There was a singular dearth of trustworthy information on the subject at our army headquarters. We moved forward by very short marches of six or eight miles, feeling our way so cautiously that Lee's reports speak of it as an unexpectedly slow approach. The Comte de Paris excuses it on the ground of the disorganized condition of McClellan's army after the recent battle. It must be remembered, however, that Sumner's corps and Franklin's had not been at the second Bull Run, and were veterans of the Potomac Army. The Twelfth Corps had been Banks's, and it too had not been engaged at the second Bull Run, its work having been to cover the trains of Pope's army on the retrograde movement from Warrenton Junction. Although new regiments had been added to these corps, it is hardly proper to say that the army as a whole was not one which could be rapidly manoeuvred. I see no good reason why it might not have advanced at once to the left bank of the Monocacy, covering thus both Washington and Baltimore, and hastening by some days Lee's movement across the Blue Ridge. We should at least have known where the enemy was by being in contact with him, instead of being the sport of all sorts of vague rumors and wild reports. [*Note: McClellan was not wholly responsible for this tardiness, for Halleck was very timid about uncovering Washington, and his dispatches tended to increase McClellan's natural indecision. Official Records, vol. xix. pt. ii. p. 280.*]

The Kanawha division took the advance of the right wing when we left Leesboro on the 8th, and marched to Brookville. On the 9th it reached Goshen, where it lay on the 10th, and on the 11th reached Ridgeville on the railroad. The rest of the Ninth Corps was an easy march behind us. Hooker had been ordered further to the right on the strength of rumors that Lee was making a circuit towards Baltimore, and his corps reached Cooksville and the railroad some ten miles east of my position. The extreme left of the army was at Poolesville, near the Potomac, making a spread of thirty miles across the whole front. The cavalry did not succeed in getting far in advance of the infantry, and very little valuable information was obtained. At Ridgeville, however, we got reliable evidence that Lee had evacuated Frederick the day before, and that only cavalry was east of the Catoctin Mountains. Hooker got similar information at about the same time. It was now determined to move more rapidly, and early in the morning of the 12th I was ordered to march to New Market and thence to Frederick. At New Market I was overtaken by General Reno, with several officers of rank from the other divisions of the corps, and they dismounted at a little tavern by the roadside to see the Kanawha division go by. Up to this time they had seen nothing of us whatever. The men had been so long in the West Virginia mountains at hard service, involving long and rapid marches, that they had much the same strength of legs and ease in marching which was afterward so much talked of when seen in Sherman's army at the review in Washington at the close of the war. I stood a little behind Reno and the rest, and had the pleasure of hearing their involuntary exclamations of admiration at the marching of the men. The easy swinging step, the graceful poise of the musket on the shoulder, as if it were a toy and not a burden, and the compactness of the column were all noticed and praised with a heartiness which was very grateful to my ears. I no longer felt any doubt that the division stood well in the opinion of my associates.

I enjoyed this the more because, the evening before, a little incident had occurred which had threatened to result in some ill-feeling. It had been thought that we were likely to be attacked at Ridgeville, and on reaching the village I disposed the division so as to cover the place and to be ready for an engagement. I ordered the brigades to bivouac in line of battle, covering the front with outposts and with cavalry vedettes from the Sixth New York Cavalry (Colonel Devin), which had been attached to the division during the advance. The men were without tents, and to make beds had helped themselves to some straw from stacks in the vicinity. Toward evening General Reno rode up, and happening first to meet Lieutenant-Colonel Rutherford B. Hayes, commanding the Twenty third Ohio, he rather sharply inquired why the troops were not bivouacking "closed in mass," and also blamed the taking of the straw. Colonel Hayes referred him to me as the proper person to account for the disposition of the troops, and quietly said he thought the quartermaster's department could settle for the straw if the owner was loyal. A few minutes later the general came to my own position, but was now quite over his irritation. I, of course, knew nothing of his interview with Hayes, and when he said that it was the policy in Maryland to make the troops bivouac in compact mass, so as to do as little damage to property as possible, I cordially assented, but urged that such a rule would not apply to the advance-guard when supposed to be in presence of the enemy; we needed to have the men already in line if an alarm should be given in the night. To this he agreed, and a pleasant conversation followed. Nothing was said to me about the straw taken for bedding, and when I heard of the little passage-at-arms with Colonel Hayes, I saw that it was a momentary disturbance which had no real significance. Camp gossip, however, is as bad as village gossip, and in a fine volume of the "History of the Twenty-first Massachusetts Regiment," I find it stated that the Kanawha division coming fresh from the West was disposed to plunder and pillage, giving an exaggerated version of the foregoing story as evidence of it. This makes it a duty to tell what was the small foundation for the charge, and to say that I believe no regiments in the army were less obnoxious to any just accusation of such a sort. The gossip would never have survived the war at all but for the fact that Colonel Hayes became President of the United States, and the supposed incident of his army life thus acquired a new interest. *[Note: This incident gives me the opportunity to say that after reading a good many regimental histories, I am struck with the fact that with the really invaluable material they contain when giving the actual experiences of the regiments themselves, they also embody a great deal of mere gossip. As a rule, their value is confined to what strictly belongs to the regiment; and the criticisms, whether of other organizations or of commanders, are likely to be the expression of the local and temporary prejudices and misconceptions which are notoriously current in time of war. They need to be read with due allowance for this. The volume referred to is a favorable example of its class, but its references to the Kanawha division (which was in the Ninth Corps only a month) illustrate the tendency I have mentioned. It should be borne in mind that the Kanawha men had the position of advance-guard, and I believe did not camp in the neighborhood of the other divisions in a single instance from the time we left Leesboro till the battle of South Mountain. What is said of them, therefore, is not from observation. The incident between Reno and Hayes occurred in the camp of the latter, and could not possibly be known to the author of the regimental history but by hearsay. Yet he affirms as a fact that the*

Kanawha division "plundered the country unmercifully," for which Reno "took Lieutenant-Colonel Hayes severely though justly to task." He also asserts that the division set a "very bad example" in straggling. As to this, the truth is as I have circumstantially stated it above. He has still further indulged in a "slant" at the "Ohioans" in a story of dead Confederates being put in a well at South Mountain,—a story as apocryphal as the others. Wise's house and well were within the camp of the division to which the Twenty-first Massachusetts belonged, and the burial party there would have been from that division. Lastly, the writer says that General Cox, the temporary corps commander, "robs us [the Twenty-first Massachusetts] of our dearly bought fame" by naming the Fifty-first New York and Fifty-first Pennsylvania as the regiments which stormed the bridge at Antietam. He acquits Burnside and McClellan of the alleged injustice, saying they "follow the corps report in this respect." Yet mention is not made of the fact that my report literally copies that of the division commander, who himself selected the regiments for the charge! The "Ohioan" had soon gone west again with his division, and was probably fair game. There is something akin to provincialism in regimental esprit de corps, and such instances as the above, which are all found within a few pages of the book referred to, show that, like Leech's famous Staffordshire rough in the Punch cartoon, to be a "stranger" is a sufficient reason to " 'eave 'arf a brick at un." See letters of President Hayes and General Crook on the subject, Appendix B.]

From New Market we sent the regiment of cavalry off to the right to cover our flank, and to investigate reports that heavy bodies of the enemy's cavalry were north of us. The infantry pushed rapidly toward Frederick. The opposition was very slight till we reached the Monocacy River, which is perhaps half a mile from the town. Here General Wade Hampton, with his brigade as rear-guard of Lee's army, attempted to resist the crossing. The highway crosses the river by a substantial stone bridge, and the ground upon our bank was considerably higher than that on the other side. We engaged the artillery of the enemy with a battery of our own, which had the advantage of position, whilst the infantry forced the crossing both by the bridge and by a ford a quarter of a mile to the right. As soon as Moor's brigade was over, it was deployed on the right and left of the turnpike, which was bordered on either side by a high and strong post-and-rail fence. Scammon's was soon over, and similarly deployed as a second line, with the Eleventh Ohio in column in the road. Moor had with him a troop of horse and a single cannon, and went forward with the first line, allowing it to keep abreast of him on right and left. I also rode on the turnpike between the two lines, and only a few rods behind Moor, having with me my staff and a few orderlies. Reno was upon the other bank of the river, overlooking the movement, which made a fine military display as the lines advanced at quick-step toward the city. Hampton's horsemen had passed out of our sight, for the straight causeway turned sharply to the left just as it entered the town, and we could not see beyond the turn. We were perhaps a quarter of a mile from the city, when a young staff officer from corps headquarters rode up beside me and exclaimed in a boisterous way, "Why don't they go in faster? There's nothing there!" I said to the young man, "Did General Reno send you with any order to me?" "No," he replied. "Then," said I, "when I want your advice I will ask it." He moved off abashed, and I did not notice what had become of him, but, in fact, he rode up to Colonel Moor, and repeated a similar speech. Moor was stung by the impertinence which he assumed to be a criticism upon him from corps headquarters, and, to my amazement, I saw him suddenly dash ahead at a gallop with his escort and the gun. He soon came to the turn of the road where it loses itself among the houses; there was a quick, sharp rattling of carbines, and Hampton's cavalry was atop of the little party. There was one discharge of the cannon, and some of the brigade staff and escort came back in disorder. I ordered up at "double quick" the Eleventh Ohio, which, as I have said, was in column in the road, and these, with bayonets fixed, dashed into the town. The enemy had not waited for them, but retreated out of the place by the Hagerstown road. Moor had been ridden down, unhorsed, and captured. The artillery-men had unlimbered the gun, pointed it, and the gunner stood with the lanyard in his hand, when he was struck by a charging horse; the gun was fired by the concussion, but at the same moment it was capsized into the ditch by the impact of the cavalry column. The enemy had no time to right the gun or carry it off, nor to stop for prisoners. They forced Moor on another horse, and turned tail as the charging lines of infantry came up on right and left as well as the column in the road, for there had not been a moment's pause in the advance. It had all happened, and the gun with a few dead and wounded of both sides were in our hands, in less time than it has taken to describe it. Those who may have a fancy for learning how Munchausen would tell this story, may find it in the narrative of Major Heros von Borke of J. E. B. Stuart's staff. [*Note: Von Borke's account is so good an example of the way in which romance may be built up out of a little fact that I give it in full. The burning of the stone bridge half a mile in rear of the little affair was a peculiarly brilliant idea; but he has evidently confused our advance with that on the Urbana road. He says: "Toward evening the enemy arrived in the immediate neighborhood of Monocacy bridge, and observing only a small force at this point, advanced very carelessly. A six-pounder gun had been placed in position by them at a very short distance from the bridge, which fired from time to time a shot at our horsemen, while the foremost regiment marched along at their ease, as if they believed this small body of cavalry would soon wheel in flight. This favorable moment for an attack was seized in splendid style by Major Butler, who commanded the two squadrons of the Second South Carolina Cavalry, stationed at this point as our rear-guard. Like lightning he darted across the bridge, taking the piece of artillery, which had scarcely an opportunity of firing a shot, and falling upon the regiment of infantry, which was dispersed in a few seconds, many of them being shot down, and many others, among whom was the colonel in command, captured. The colors of the regiment also fell into Major Butler's hands. The piece of artillery, in the hurry of the moment, could not be brought over to our side of the river, as the enemy instantly sent forward a large body of cavalry at a gallop, and our dashing men had only time to spike it and trot with their prisoners across the bridge, which, having been already fully prepared for burning, was in a blaze when the infuriated Yankees arrived at the water's edge. The conflagration of the bridge of course checked their onward movement, and we quietly continued the retreat." Von Borke, vol. i. p. 203. Stuart's report is very nearly accurate: Official Records, vol. xix. pt. i. p. 816.]* Moor's capture, however, had consequences, as we shall see. The command of his brigade passed to Colonel George Crook of the Thirty-sixth Ohio.

Frederick was a loyal city, and as Hampton's cavalry went out at one end of the street and our infantry came in at the other, and whilst the carbine smoke and the smell of powder still lingered, the closed window-shutters of the houses flew open, the sashes went up, the windows were filled with ladies waving their handkerchiefs and national flags, whilst the men came to the column with fruits and refreshments for the marching soldiers as they went by in the hot sunshine of the September afternoon. *[Note: Although at the head of the column, the "truth of history" compels me to say that I saw nothing of Barbara Frietchie, and heard nothing of her till I read Whittier's poem in later years. When, however, I visited Frederick with General Grant in 1869, we were both presented with walking-sticks made from timbers of Barbara's house which had been torn down, and, of course, I cannot dispute the story of which I have the stick as evidence; for Grant thought the stick shut me up from any denial and established the legend.]* Pleasonton's cavalry came in soon after by the Urbana road, and during the evening a large part of the army drew near the place. Next morning (13th) the cavalry went forward to reconnoitre the passes of Catoctin Mountain, Rodman's division of our corps being ordered to support them and to proceed toward Middletown in the Catoctin valley. Through some misunderstanding Rodman took the road to Jefferson, leading to the left, where Franklin's corps was moving, and did not get upon the Hagerstown road. About noon I was ordered to march upon the latter road to Middletown. McClellan himself met me as my column moved out of town, and told me of the misunderstanding in Rodman's orders, adding that if I found him on the march I should take his division also along with me. *[Note: As is usual in such cases, the direction was later put in writing by his chief of staff. Official Records, vol. li. pt. i. p. 827.]* I did not meet him, but the other two divisions of the corps crossed Catoctin Mountain that night, whilst Rodman returned to Frederick. The Kanawha division made an easy march, and as the cavalry was now ahead of us, met no opposition in crossing Catoctin Mountain or in the valley beyond. On the way we passed a house belonging to a branch of the Washington family, and a few officers of the division accompanied me, at the invitation of the occupant, to look at some relics of the Father of his Country which were preserved there. We stood for some minutes with uncovered heads before a case containing a uniform he had worn, and other articles of personal use hallowed by their association with him, and went on our way with our zeal strengthened by closer contact with souvenirs of the great patriot. Willcox's division followed us, and encamped a mile and a half east of Middletown. Sturgis's halted not far from the western foot of the mountain, with corps headquarters near by. My own camp for the night was pitched in front (west) of the village of Middletown along Catoctin Creek. Pleasonton's cavalry was a little in advance of us, at the forks of the road where the old Sharpsburg road turns off to the left from the turnpike. The rest of the army was camped about Frederick, except Franklin's corps (Sixth), which was near Jefferson, ten miles further south but also east of Catoctin Mountain.

The Catoctin or Middletown valley is beautifully included between Catoctin Mountain and South Mountain, two ranges of the Blue Ridge, running northeast and southwest. It is six or eight miles wide, watered by Catoctin Creek, which winds southward among rich farms and enters the Potomac near Point of Rocks. The National road leaving Frederick passes through Middletown and crosses South Mountain, as it goes northwestward, at a depression called Turner's Gap. The old Sharpsburg road crosses the summit at another gap, known as Fox's, about a mile south of Turner's. Still another, the old Hagerstown road, finds a passage over the ridge at about an equal distance north. The National road, being of easier grades and better engineering, was now the principal route, the others having degenerated to rough country roads. The mountain crests are from ten to thirteen hundred feet above the Catoctin valley, and the "gaps" are from two to three hundred feet lower than the summits near them. *[Note: These elevations are from the official map of the U.S. Engineers.]* These summits are like scattered and irregular hills upon the high rounded surface of the mountain top. They are wooded, but along the southeasterly slopes, quite near the top of the mountain, are small farms, with meadows and cultivated fields.

The military situation had been cleared up by the knowledge of Lee's movements which McClellan got from a copy of Lee's order of the day for the both. This had been found at Frederick on the 13th, and it tallied so well with what was otherwise known that no doubt was left as to its authenticity. It showed that Jackson's corps with Walker's division were besieging Harper's Ferry on the Virginia side of the Potomac, whilst McLaws's division supported by Anderson's was co-operating on Maryland Heights. *[Note: Official Records, vol. xix. pt. ii. pp. 281, 603.]* Longstreet, with the remainder of his corps, was at Boonsboro or near Hagerstown. D. H. Hill's division was the rear-guard, and the cavalry under Stuart covered the whole, a detached squadron being with Longstreet, Jackson, and McLaws each. The order did not name the three separate divisions in Jackson's command proper (exclusive of Walker), nor those remaining with Longstreet except D. H. Hill's; but it is hardly conceivable that these were not known to McClellan after his own and Pope's contact with them during the campaigns of the spring and summer. At any rate, the order showed that Lee's army was in two parts, separated by the Potomac and thirty or forty miles of road. As soon as Jackson should reduce Harper's Ferry they would reunite. Friday the 12th was the day fixed for the concentration of Jackson's force for his attack, and it was Saturday when the order fell into McClellan's hands. Three days had already been lost in the slow advance since Lee had crossed Catoctin Mountain, and Jackson's artillery was now heard pounding at the camp and earthworks of Harper's Ferry. McLaws had already driven our forces from Maryland Heights, and had opened upon the ferry with his guns in commanding position on the north of the Potomac. *[Note: Id., p. 607.]* McClellan telegraphed to the President that he would catch the rebels "in their own trap if my men are equal to the emergency." *[Note: Official Records, vol. xix. pt. ii. p. 281.]* There was certainly no time to lose. The information was in his hands before noon, for he refers to it in a dispatch to Mr. Lincoln at twelve. If his men had been ordered to be at the top of South Mountain before dark, they could have been there; but less than one full corps passed Catoctin Mountain that day or night, and when the leisurely movement of the 14th began, he himself, instead of being with the advance, was in Frederick till after 2 P.M., at which hour he sent a dispatch to Washington, and then rode to the front ten or twelve miles away. The failure to be "equal to the emergency" was not in his men. Twenty-four hours, as it turned out, was the whole difference between saving and losing Harper's Ferry with its ten or twelve thousand men and its unestimated munitions and stores. It may be that the commanders of the garrison were in fault, and that a more stubborn resistance should have been made. It may be that Halleck ought to have ordered the place to be evacuated earlier, as McClellan suggested. Nevertheless, at noon of the 13th McClellan had it in his power to save the place and interpose his army between the two wings, of the Confederates with decisive effect on the campaign. He saw that it was an "emergency," but did not call upon his men for any extraordinary exertion. Harper's. Ferry surrendered, and Lee united the wings of his army beyond the Antietam before the final and general engagement was forced upon him.

At my camp in front of Middletown, I received no orders looking to a general advance on the 14th; but only to support, by a detachment, Pleasonton's cavalry in a reconnoissance toward Turner's Gap.

Pleasonton himself came to my tent in the evening, and asked that one brigade might report to him in the morning for the purpose. Six o'clock was the hour at which he wished them to march. He said further that he and Colonel Crook were old army acquaintances and that he would like Crook to have the detail. I wished to please him, and not thinking that it would make any difference to my brigade commanders, intimated that I would do so. But Colonel Scammon, learning what was intended, protested that under our custom his brigade was entitled to the advance next day, as the brigades had taken it in turn. I explained that it was only as a courtesy to Pleasonton and at his request that the change was proposed. This did not better the matter in Scammon's opinion. He had been himself a regular officer, and the point of professional honor touched him. I recognized the justice of his demand, and said he should have the duty if he insisted upon it. Pleasonton was still in the camp visiting with Colonel Crook, and I explained to him the reasons why I could not yield to his wish, but must assign Scammon's brigade to the duty in conformity with the usual course. There was in fact no reason except the personal one for choosing one brigade more than the other, for they were equally good. Crook took the decision in good part, though it was natural that he should wish for an opportunity of distinguished service, as he had not been the regular commandant of the brigade. Pleasonton was a little chafed, and even intimated that he claimed some right to name the officer and command to be detailed. This, of course, I could not admit, and issued the formal orders at once. The little controversy had put Scammon and his whole brigade upon their mettle, and was a case in which a generous emulation did no harm. What happened in the morning only increased their spirit and prepared them the better to perform what I have always regarded as a very brilliant exploit.

The morning of Sunday the 14th of September was a bright one. I had my breakfast very early and was in the saddle before it was time for Scammon to move. He was prompt, and I rode on with him to see in what way his support was likely to be used. Two of the Ninth Corps batteries (Gibson's and Benjamin's) had accompanied the cavalry, and one of these was a heavy one of twenty-pounder Parrotts. They were placed upon a knoll a little in front of the cavalry camp, about half a mile beyond the forks of the old Sharpsburg road with the turnpike. They were

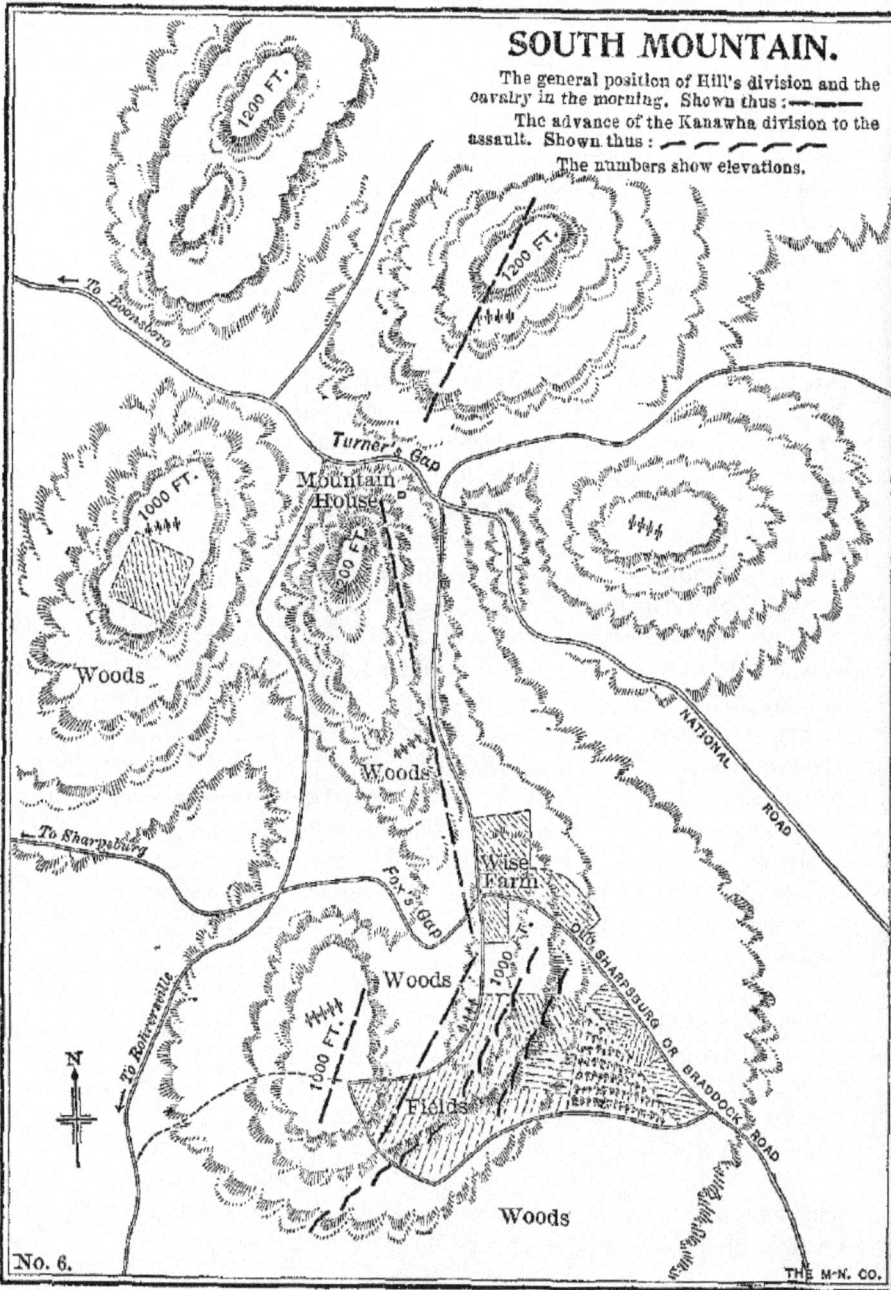

SOUTH MOUNTAIN.

The general position of Hill's division and the cavalry in the morning. Shown thus:

The advance of the Kanawha division to the assault. Shown thus:

The numbers show elevations.

exchanging shots with a battery of the enemy well up in the gap. Just as Scammon and I crossed Catoctin Creek I was surprised to see Colonel Moor standing at the roadside. With astonishment I rode to him and asked how he came there. He said that he had been taken beyond the mountain after his capture, but had been paroled the evening before, and was now finding his way back to us on foot. "But where are you going?" said he. I answered that Scammon was going to support Pleasonton in a reconnoissance into the gap. Moor made an involuntary start, saying, "My God! be careful!" then checking himself, added, "But I am paroled!" and turned away. I galloped to Scammon and told him that I should follow him in close support with Crook's brigade, and as I went back along the column I spoke to each regimental commander, warning them to be prepared for anything, big or little,—it might be a skirmish, it might be a battle. Hurrying to camp, I ordered Crook to turn out his brigade and march at once. I then wrote a dispatch to General Reno, saying I suspected we should find the enemy in force on the mountain top, and should go forward with both brigades instead of sending one. Starting a courier with this, I rode forward again and found Pleasonton. Scammon had given him an inkling of our suspicions, and in the personal interview they had reached a mutual good understanding. I found that he was convinced that it would be unwise to make an attack in front, and had determined that his horsemen should merely demonstrate upon the main road and support the batteries, whilst Scammon should march by the old Sharpsburg road and try to reach the flank of the force on the summit. I told him that in view of my fear that the force of the enemy might be too great for Scammon, I had determined to bring forward Crook's brigade in support. If it became necessary to fight with the whole division, I should do so, and in that case I should assume the responsibility myself as his senior officer. To this he cordially assented.

One section of McMullin's six-gun battery was all that went forward with Scammon (and even these not till the infantry reached the summit), four guns being left behind, as the road was rough and steep. There were in Simmonds's battery two twenty-pounder Parrott guns, and I ordered these also to remain on the turnpike and to go into action with Benjamin's battery of the same calibre. It was about half-past seven when Crook's head of column filed off from the turnpike upon the old Sharpsburg road, and Scammon had perhaps half an hour's start. We had fully two miles to go before we should reach the place where our attack was actually made, and as it was a pretty sharp ascent the men marched slowly with frequent rests. On our way up we were overtaken by my courier who had returned from General Reno with approval of my action and the assurance that the rest of the Ninth Corps would come forward to my support.

When Scammon had got within half a mile of Fox's Gap (the summit of the old Sharpsburg road), *[Note: The Sharpsburg road is also called the Braddock road, as it was the way by which Braddock and Washington had marched to Fort Duquesne (Pittsburg) in the old French war. For the same reason the gap is called Braddock's Gap. I have adopted that which seems to be in most common local use.]* the enemy opened upon him with case-shot from the edge of the timber above the open fields, and he had judiciously turned off upon a country road leading still further to the left, and nearly parallel to the ridge above. His movement had been made under cover of the forest, and he had reached the extreme southern limit of the open fields south of the gap on this face of the mountain. Here I overtook him, his brigade being formed in line under cover of the timber, facing open pasture fields having a stone wall along the upper side, with the forest again beyond this. On his left was the Twenty-third Ohio under Lieutenant-Colonel R. B. Hayes, who had been directed to keep in the woods beyond the open, and to strike if possible the flank of the enemy. His centre was the Twelfth Ohio under Colonel Carr B. White, whose duty was to attack the stone wall in front, charging over the broad open fields. On the right was the Thirtieth Ohio, Colonel Hugh Ewing, who was ordered to advance against a battery on the crest which kept up a rapid and annoying fire. It was now about nine o'clock, and Crook's column had come into close support. Bayonets were fixed, and at the word the line rushed forward with loud hurrahs. Hayes, being in the woods, was not seen till he had passed over the crest and turned upon the enemy's flank and rear. Here was a sharp combat, but our men established themselves upon the summit and drove the enemy before them. White and Ewing charged over the open under a destructive fire of musketry and shrapnel. As Ewing approached the enemy's battery (Bondurant's), it gave him a parting salvo, and limbered rapidly toward the right along a road in the edge of the woods which follows the summit to the turnpike near the Mountain House at Turner's Gap. White's men never flinched, and the North Carolinians of Garland's brigade (for it was they who held the ridge at this point) poured in their fire till the advancing line of bayonets was in their faces when they broke away from the wall. Our men fell fast, but they kept up their pace, and the enemy's centre was broken by a heroic charge. Garland strove hard to rally his men, but his brigade was hopelessly broken in two. He rallied his right wing on the second ridge a little in rear of that part of his line, but Hayes's regiment was here pushing forward from our left. Colonel Ruffin of the Thirteenth North Carolina held on to the ridge road beyond our right, near Fox's Gap. The fighting was now wholly in the woods, and though the enemy's centre was routed there was stubborn resistance on both flanks. His cavalry dismounted (said to be under Colonel Rosser *[Note: Stuart's Report, Official Records, vol. xix. pt. i. p. 817.]*) was found to extend beyond Hayes's line, and supported the Stuart artillery, which poured canister into our advancing troops. I now ordered Crook to send the Eleventh Ohio (under Lieutenant-Colonel Coleman) beyond Hayes's left to extend our line in that direction, and to direct the Thirty-sixth Ohio (Lieutenant-Colonel Clark) to fill a gap between the Twelfth and Thirtieth caused by diverging lines of advance. The only remaining regiment (the Twenty-eighth, Lieutenant-Colonel Becker) was held in reserve on the right. The Thirty-sixth aided by the Twelfth repulsed a stout effort of the enemy to re-establish their centre. The whole line again sprung forward. A high knoll on our left was carried. The dismounted cavalry was forced to retreat with their battery across the ravine in which the Sharpsburg road descends on the west of the mountain, and took a new position on a separate hill in rear of the heights at the Mountain House. There was considerable open ground at this new position, from which their battery had full play at a range of about twelve hundred yards upon the ridge held by us. But the Eleventh and Twenty-third stuck stoutly to the hill which Hayes had first carried, and their line was nearly parallel to the Sharpsburg road, facing north. Garland had rushed to the right of his brigade to rally them when they had broken before the onset of the Twenty-third Ohio upon the flank, and in the desperate contest there he had been killed and the disaster to his command made irreparable. On our side Colonel Hayes had also been disabled by a severe wound as he gallantly led the Ohio regiment.

I now directed the centre and right to push forward toward Fox's Gap. Lieutenant Croome with a section of McMullin's battery had come up, and he put his guns in action in the most gallant manner in the open ground near Wise's house. The Thirtieth and Thirty-sixth changed front to the right and attacked the remnant of Garland's brigade, now commanded by Colonel McRae, and drove it and two regiments from G. B. Anderson's brigade back upon the wooded hill beyond Wise's farm at Fox's Gap. The whole of Anderson's brigade retreated further along the crest toward the Mountain House. Meanwhile the Twelfth Ohio, also changing front, had thridded its way in the same direction through laurel thickets on the reverse slope of the mountain, and attacking suddenly the force at Wise's as the other two regiments charged it in front, completed the rout and brought off two hundred prisoners. Bondurant's battery was again driven hurriedly off to the north. But the hollow at the gap about Wise's was no place to stay. It was open ground and was swept by the batteries of the cavalry on the open hill to the northwest, and by those of Hill's division about the Mountain House and upon the highlands north of the National road; for those hills run forward like a bastion and give a perfect flanking fire along our part of the mountain. The gallant Croome with a number of his gunners had been killed, and his guns were brought back into the shelter of the woods, on the hither side of Wise's fields. The infantry of the right wing was brought to the same position, and our lines were reformed along the curving crests from that point which looks down into the gap and the Sharpsburg road, toward the left. The extreme right with Croome's two guns was held by the Thirtieth, with the Twenty-eighth in second line. Next came the Twelfth, with the Thirty-sixth in second line, the front curving toward the west with the form of the mountain summit. The left of the Twelfth dipped a little into a hollow, beyond which the Twenty-third and Eleventh occupied the next hill facing toward the Sharpsburg road. Our front was hollow, for the two wings were nearly at right angles to each other; but the flanks were strongly placed, the right, which was most exposed, having open ground in front which it could sweep with its fire and having the reserve regiments closely supporting it. Part of Simmonds's battery which had also come up had done good service in the last combats, and was now disposed so as to check the fire of the enemy.

It was time to rest. Three hours of up-hill marching and climbing had been followed by as long a period of bloody battle, and it was almost noon. The troops began to feel the exhaustion of such labor and struggle. We had several hundred prisoners in our hands, and the field was thickly strewn with dead, in gray and in blue, while our field hospital a little down the mountain side was encumbered with hundreds of wounded. We learned from our prisoners that the summit was held by D. H. Hill's division of five brigades with Stuart's cavalry, and that Longstreet's corps was in close support. I was momentarily expecting to hear from the supporting divisions of the Ninth Corps, and thought it the part of wisdom to hold fast to our strong position astride of the mountain top commanding the Sharpsburg road till our force should be increased. The two Kanawha brigades had certainly won a glorious victory, and had made so assured a success of the day's work that it would be folly to imperil it. *[Note: For Official Records, see Official Records, vol. xix. pt. i. pp. 458-474.]*

General Hill has since argued that only part of his division could oppose us; *[Note: Century War Book, vol. ii. pp. 559, etc.]* but his brigades were all on the mountain summit within easy support of each other, and they had the day before them. It was five hours from the time of our first charge to the arrival of our first supports, and it was not till three o'clock in the afternoon that Hooker's corps reached the eastern base of the mountain and began its deployment north of the National road. Our effort was to attack the weak end of his line, and we succeeded in putting a stronger force there than that which opposed us. It is for our opponent to explain how we were permitted to do it. The two brigades of the Kanawha division numbered less than 3000 men. Hill's division was 5000 strong, *[Note: Official Records, vol. xix. pt. i. p. 1025.]* even by the Confederate method of counting their effectives, which should be increased nearly one-fifth to compare properly with our reports. In addition to these Stuart had the principal part of the Confederate cavalry on this line, and they were not idle spectators. Parts of Lee's and Hampton's brigades were certainly there, and probably the whole of Lee's. *[Note:Id., p. 819.]* With less than half the numerical strength which was opposed to it, therefore, the Kanawha division had carried the summit, advancing to the charge for the most part over open ground in the storm of musketry and artillery fire, and held the crests they had gained through the livelong day, in spite of all efforts to retake them.

In our mountain camps of West Virginia I had felt discontented that our native Ohio regiments did not take as kindly to the labors of drill and camp police as some of German birth, and I had warned them that they would feel the need of accuracy and mechanical precision when the day of battle came. They had done reasonably well, but suffered in comparison with some of the others on dress parade and in the form and neatness of the camp. When, however, on the slopes of South Mountain I saw the lines go forward steadier and more even under fire than they ever had done at drill, their intelligence making them perfectly comprehend the advantage of unity in their effort and in the shock when they met the foe—when their bodies seemed to dilate, their step to have better cadence and a tread as of giants as they went cheering up the hill,—I took back all my criticisms and felt a pride and glory in them as soldiers and comrades that words cannot express.

It was about noon that the lull in the battle occurred, and it lasted a couple of hours, while reinforcements were approaching the mountain top from both sides. The enemy's artillery kept up a pretty steady fire, answered occasionally by our few cannon; but the infantry rested on their arms, the front covered by a watchful line of skirmishers, every man at his tree. The Confederate guns had so perfectly the range of the sloping fields about and behind us, that their canister shot made long furrows in the sod with a noise like the cutting of a melon rind, and the shells which skimmed the crest and burst in the tree-tops at the lower side of the fields made a sound like the crashing and falling of some brittle substance, instead of the tough fibre of oak and pine. We had time to notice these things as we paced the lines waiting for the renewal of the battle.

Willcox's division reported to me about two o'clock, and would have been up earlier, but for a mistake in the delivery of a message to him. He had sent from Middletown to ask me where I desired him to come, and finding that the messenger had no clear idea of the roads by which he had travelled, I directed him to say that General Pleasonton would point out the road I had followed, if inquired of. Willcox understood the messenger that I wished him to inquire of Pleasonton where he had better put his division in, and on doing so, the latter suggested that he move against the crests on the north of the National road. He was preparing to do this when Burnside and Reno came up and corrected the movement, recalling him from the north and sending him by the old Sharpsburg road to my position. As his head of column came up, Longstreet's corps was already forming with its right outflanking my left. I sent two regiments *[Note: In my official report I said one regiment, but*

General Willcox reported that he sent two, and he is doubtless right. For his official report, see Official Records, vol. xix. pt. ii. p. 428.] to extend my left, and requested Willcox to form the rest of the division on my right facing the summit. He was doing this when he received an order from General Reno to take position overlooking the National road facing northward. *[Note: Ibid.]* I can hardly think the order could have been intended to effect this, as the turnpike is deep between the hills there, and the enemy quite distant on the other side of the gorge. But Willcox, obeying the order as he received it, formed along the Sharpsburg road, his left next to my right, but his line drawn back nearly at right angles to it. He placed Cook's battery in the angle, and this opened a rapid fire on one of the enemy's which was on the bastion-like hill north of the gorge already mentioned. Longstreet's men were now pretty well up, and pushed a battery forward to the edge of the timber beyond Wise's farm, and opened upon Willcox's line, enfilading it badly. There was a momentary break there, but Willcox was able to check the confusion, and to reform his lines facing westward as I had originally directed; Welch's brigade was on my right, closely supporting Cook's battery and Christ's beyond it. The general line of Willcox's division was at the eastern edge of the wood looking into the open ground at Fox's Gap, on the north side of the Sharpsburg road. A warm skirmishing fight was continued along the whole of our line, our purpose being to hold fast my extreme left which was well advanced upon and over the mountain crest, and to swing the right up to the continuation of the same line of hills near the Mountain House.

At nearly four o'clock the head of Sturgis's column approached. *[Note: Sturgis's Report, Id., pt. i. p. 443.]* McClellan had arrived on the field, and he with Burnside and Reno was at Pleasonton's position at the knoll in the valley, and from that point, a central one in the midst of the curving hills, they issued their orders. They could see the firing of the enemy's battery from the woods beyond the open ground in front of Willcox, and sent orders to him to take or silence those guns at all hazards. He was preparing to advance, when the Confederates anticipated him (for their formation had now been completed) and came charging out of the woods across the open fields. It was part of their general advance and their most determined effort to drive us from the summit we had gained in the morning. The brigades of Hood, Whiting, Drayton, and D. R. Jones in addition to Hill's division (eight brigades in all) joined in the attack on our side of the National road, batteries being put in every available position. *[Note: Longstreet's Report, Official Records, vol. xix. pt. i. p. 839.]* The fight raged fiercely along the whole front, but the bloodiest struggle was around Wise's house, where Drayton's brigade assaulted my right and Willcox's left, coming across the open ground. Here the Sharpsburg road curves around the hill held by us so that for a little way it was parallel to our position. As the enemy came down the hill forming the other side of the gap, across the road and up again to our line, they were met by so withering a fire that they were checked quickly, and even drifted more to the right where their descent was continuous. Here Willcox's line volleyed into them a destructive fire, followed by a charge that swept them in confusion back along the road, where the men of the Kanawha division took up the attack and completed their rout. Willcox succeeded in getting a foothold on the further side of the open ground and driving off the artillery which was there. Along our centre and left where the forest was thick, the enemy was equally repulsed, but the cover of the timber enabled them to keep a footing near by, whilst they continually tried to extend so as to outflank us, moving their troops along a road which goes diagonally down that side of the mountain from Turner's Gap to Rohrersville. The batteries on the north of the National road had been annoying to Willcox's men as they advanced, but Sturgis sent forward Durell's battery from his division as soon as he came up, and this gave special attention to these hostile guns, diverting their fire from the infantry. Hooker's men, of the First Corps, were also by this time pushing up the mountain on that side of the turnpike, and we were not again troubled by artillery on our right flank.

It was nearly five o'clock when the enemy had disappeared in the woods beyond Fox's Gap and Willcox could reform his shattered lines. As the easiest mode of getting Sturgis's fresh men into position, Willcox made room on his left for Ferrero's brigade supported by Nagle's, doubling also his lines at the extreme right. Rodman's division, the last of the corps, now began to reach the summit, and as the report came from the extreme left that the enemy was stretching beyond our flank, I sent Fairchild's brigade to assist our men there, whilst Rodman took Harland's to the support of Willcox. A staff officer now brought word that McClellan directed the whole line to advance. At the left this could only mean to clear our front decisively of the enemy there, for the slopes went steadily down to the Rohrersville road. At the centre and right, whilst we held Fox's Gap, the high and rocky summit at the Mountain House was still in the enemy's possession. The order came to me as senior officer upon the line, and the signal was given. On the left Longstreet's men were pushed down the mountain side beyond the Rohrersville and Sharpsburg roads, and the contest there was ended. The two hills between the latter road and the turnpike were still held by the enemy, and the further one could not be reached till the Mountain House should be in our hands. Sturgis and Willcox, supported by Rodman, again pushed forward, but whilst they made progress they were baffled by a stubborn and concentrated resistance.

Reno had followed Rodman's division up the mountain, and came to me a little before sunset, anxious to know why the right could not get forward quite to the summit. I explained that the ground there was very rough and rocky, a fortress in itself and evidently very strongly held. He passed on to Sturgis, and it seemed to me he was hardly gone before he was brought back upon a stretcher, dead. He had gone to the skirmish line to examine for himself the situation, and had been shot down by the enemy posted among the rocks and trees. There was more or less firing on that part of the field till late in the evening, but when morning dawned the Confederates had abandoned the last foothold above Turner's Gap and retreated by way of Boonsboro to Sharpsburg. The casualties in the Ninth Corps had been 889, of which 356 were in the Kanawha division. Some 600 of the enemy were captured by my division and sent to the rear under guard.

On the north of the National road the First Corps under Hooker had been opposed by one of Hill's brigades and four of Longstreet's, and had gradually worked its way along the old Hagerstown road, crowning the heights in that direction after dark in the evening. Gibbon's brigade had also advanced in the National road, crowding up quite close to Turner's Gap and engaging the enemy in a lively combat. It is not my purpose to give a detailed history of events which did not come under my own eye. It is due to General Burnside, however, to note Hooker's conduct toward his immediate superior and his characteristic efforts to grasp all the glory of the battle at the expense of truth and of honorable dealing with his commander and his comrades. Hooker's official report for the battle of South Mountain was dated at Washington, November 17th, when

Burnside was in command of the Army of the Potomac, and when the intrigues of the former to obtain the command for himself were notorious and near their final success. In it he studiously avoided any recognition of orders or directions received from Burnside, and ignores his staff, whilst he assumes that his orders came directly from McClellan and compliments the staff officers of the latter, as if they had been the only means of communication. This was not only insolent but a military offence, had Burnside chosen to prosecute it. He also asserts that the troops on our part of the line had been defeated and were at the turnpike at the base of the mountain in retreat when he went forward. At the close of his report, after declaring that "the forcing of the passage of South Mountain will be classed among the most brilliant and satisfactory achievements of this army," he adds, "its principal glory will be awarded to the First Corps." *[Note: Official Records, vol. xix. pt. i. pp. 214-215.]*

Nothing is more justly odious in military conduct than embodying slanders against other commands in an official report. It puts into the official records misrepresentations which cannot be met because they are unknown, and it is a mere accident if those who know the truth are able to neutralize their effect. In most cases it will be too late to counteract the mischief when those most interested learn of the slanders. All this is well illustrated in the present case. Hooker's report got on file months after the battle, and it was not till the January following that Burnside gave it his attention. I believe that none of the division commanders of the Ninth Corps learned of it till long afterward. I certainly did not till 1887, a quarter of a century after the battle, when the volume of the official records containing it was published. Burnside had asked to be relieved of the command of the Army of the Potomac after the battle of Fredericksburg unless Hooker among others was punished for insubordination. As in the preceding August, the popular sentiment of that army as an organization was again, in Mr. Lincoln's estimation, too potent a factor to be opposed, and the result was the superseding of Burnside by Hooker himself, though the President declared in the letter accompanying the appointment that the latter's conduct had been blameworthy. It was under these circumstances that Burnside learned of the false statements in Hooker's report of South Mountain, and put upon file his stinging response to it. His explicit statement of the facts will settle that question among all who know the reputation of the men, and though unprincipled ambition was for a time successful, that time was so short and things were "set even" so soon that the ultimate result is one that lovers of justice may find comfort in.

[Note: The text of Burnside's supplemental report is as follows:—

"When I sent in my report of the part taken by my command in the battle of South Mountain, General Hooker, who commanded one of the corps of my command (the right wing), had not sent in his report, but it has since been sent to me. I at first determined to pass over its inaccuracies as harmless, or rather as harming only their author; but upon reflection I have felt it my duty to notice two gross misstatements made with reference to the commands of Generals Reno and Cox, the former officer having been killed on that day, and the latter now removed with his command to the West.

"General Hooker says that as he came up to the front, Cox's corps was retiring from the contest. This is untrue. General Cox did not command a corps, but a division; and that division was in action, fighting most gallantly, long before General Hooker came up, and remained in the action all day, never leaving the field for one moment. He also says that he discovered that the attack by General Reno's corps was without sequence. This is also untrue, and when said of an officer who so nobly fought and died on that same field, it partakes of something worse than untruthfulness. Every officer present who knew anything of the battle knows that Reno performed a most important part in the battle, his corps driving the enemy from the heights on one side of the main pike, whilst that of General Hooker drove them from the heights on the other side.

"General Hooker should remember that I had to order him four separate times to move his command into action, and that I had to myself order his leading division (Meade's) to start before he would go." Official Records, vol. xix. pt. i. p. 422.]

The men of the First Corps and its officers did their duty nobly on that as on many another field, and the only spot on the honor of the day is made by the personal unscrupulousness and vainglory of its commander.

Franklin's corps had attacked and carried the ridge about five miles further south, at Crampton's Gap, where the pass had been so stubbornly defended by Mahone's and Cobb's brigades with artillery and a detachment of Hampton's cavalry as to cause considerable loss to our troops. The principal fighting was at a stone wall near the eastern base of the mountain, and when the enemy was routed from this position, he made no successful rally and the summit was gained without much more fighting. The attack at the stone wall not far from Burkettsville was made at about three o'clock in the afternoon. The Sixth Corps rested upon the summit at night.

R. B. Hayes.

The following account is taken from:

THE LIFE
PUBLIC SERVICES AND SELECT SPEECHES
OF
RUTHERFORD B. HAYES
by
J. Q. HOWARD
1876

Rutherford B. Hayes

CHAPTER IV.

TOP IN THE FIELD.

Appointed Major—Judge Advocate—Lieutenant-Colonel—South Mountain—Wounded—Fighting while Down—After Morgan—Battle of Cloyd Mountain—Charge up the Mountain—Enemy's Works Carried by Storm—First Battle of Winchester—Berryville.

That a loyal citizen of the antecedents, ardent patriotism, and impulsive nature of Rutherford B. Hayes would enter the army in the war for the Union, was to be looked for as a thing of course. He had been in the habit of obeying every call of duty, and could not therefore disobey when duty called loudest. He regarded the war waged for the supremacy of the constitution and the laws as a just and necessary war, and preferred to go into it if he knew he "was to die or be killed in the course of it." He had been a most earnest advocate of the election of Mr. Lincoln to the Presidency, and had been an anti-slavery man of established convictions long before the candidacy of Fremont for the Presidency. He did not think the Union should be destroyed to make slavery perpetual. He desired to mitigate and finally eradicate that evil. He had prayed for the election of General Harrison for the sake of the country; he had cast his first vote for Henry Clay, his second for General Taylor, and his third for General Scott. But the old Whig party having ceased to be a living organization, he gave his whole heart to the Republican party and its cause, and by political speeches, and in other ways, helped forward the movement in favor of equality of rights and laws. The insult to the flag at Fort Sumter aroused to the intensest pitch the patriotic indignation of a united North. At a great mass-meeting held in Cincinnati, R. B. Hayes was selected to give expression to the loyal voice, by being made chairman of the public committee on resolutions. It is not needful to add that these resolutions had all the fire and intensity of the popular feeling. The knowledge that it was his purpose to enter the Union army having reached Governor Dennison, that officer appointed Hayes major of the Twenty-third Ohio Volunteer Infantry, June

7, 1861. With this appointment was coupled the appointments of W. S. Rosecrans as colonel, and Stanley Matthews as lieutenant-colonel of the same regiment. Colonel Rosecrans, with the other field-officers, had just set to work organizing the new regiment, when Rosecrans was appointed brigadier-general, and ordered to take command of the Ohio troops moving in the direction of Western Virginia. Upon the promotion of Rosecrans, Colonel E. P. Scammon, an officer of military education, was placed in command of the Twenty-third.

After a brief period of discipline at Camp Chase the regiment was ordered, on the 25th of July, to Clarksburgh, West Virginia, and on the 29th went into camp at Weston. We shall not follow it in this or in subsequent campaigns, in its marching, scouting, skirmishing, or counter-marching. It is enough to say, that in this first campaign it assisted in clearing the whole mountainous region of Western Virginia of a formidable enemy.

Major Hayes was appointed by General Rosecrans, on the 19th of September, 1861, judge advocate of the department of Ohio, the duties of which service he discharged about two months. He received his first promotion, to the rank of lieutenant-colonel, October 24, 1861. Passing over less important events, we come to the first serious battle in which he was engaged.

THE BATTLE OF SOUTH MOUNTAIN
Was fought on Sunday, September 14, 1862, a beautiful, bright September day. The enemy were in possession of the crest of the mountain, where the old National road crossed it. The army of McClellan, with Burnside in advance, were pressing up that mountain by the National road as its center. General Cox's division of Burnside's corps was in advance. The brigade to which Lieutenant-colonel Hayes was attached was in advance of the division. His regiment was in advance of the brigade. He was ordered to pass up a mountain path on the left of the National road and feel for the enemy, advancing until he struck him; to push him up the mountain if he could; in short, to open the engagement. Lieutenant-colonel Hayes pushed into the woods, came upon the enemy's pickets, received their fire, and drove them in. He soon saw a strong force of the enemy coming toward the line of his advance from a neighboring hill, and went to meet them. Hayes charged into that force

with a regimental yell, and, after a fierce fight, drove them out of the woods in which he found them, into an open field near the summit. He then drove them across the field, losing many men and capturing and killing many of the enemy.

Hayes, having just given the command for a third charge, felt a stunning blow, and found that a large musket ball had struck his left arm above the elbow, carrying away and badly fracturing the entire bone. Fearing an artery might be severed, he asked a soldier to bandage his arm above the elbow, and a few minutes after, through exhaustion, he fell. Recovering from a state of unconsciousness while down, in a few moments, and observing that his men had fallen back to the woods for shelter, he sprang to his feet, and, with unusual vehemence, ordered them to come forward, which they did. He continued fighting some time at the head of his men; but falling a second time, from exhausted strength, he kept on giving orders, while down, to fight it out.

Major Comly, the second in command, then came to him to learn the orders under which the regiment was fighting, and deeming it best to assume command, owing to the critical condition of Lieutenant-colonel Hayes, gave orders that the wounded hero should be carried from the field. In an almost illegible narrative, written with the left hand just after the battle, we find this modest record, by the intrepid sufferer in this event: "While I was down I had considerable talk with a wounded Confederate lying near me. I gave him messages for my wife and friends in case I should not get up. We were right jolly and friendly. It was by no means an unpleasant experience."

The enemy in this action continued to pour a most destructive fire of musketry, grape, and canister into the Union ranks. Lieutenant-colonel Hayes again made his appearance on the field with his wound half dressed, and fought until carried off. Soon after, the rest of the brigade coming up, a brilliant bayonet charge up the hill dislodged the enemy and drove him into the woods beyond. The Twenty-third regiment in this engagement lost within eight men of half the entire force engaged.

South Mountain is inscribed on all the standards of this gallant regiment, and surrounds with a sad halo of glory the names of the living and the graves of the dead.

At the time this battle was fought, Lieutenant-Colonel Hayes was not under pay, having been mustered out of the Twenty-third regiment to take command of the Seventy-ninth. His wound preventing him from becoming colonel of the Seventy-ninth, he was, on the 24th of October, 1862, appointed colonel of his own regiment, vice Scammon, promoted. It was while at home recovering from his wounds that his wealthy uncle, Sardis Birchard, urged Colonel Hayes, to whom he was devotedly attached, to leave the army, on the ground that he had done his share, promising to himself and family abundant support; but he would not listen to the suggestion, and before his wounds were healed went back.

The following account is taken from:

THREE YEARS IN THE SIXTH CORPS.
A CONCISE NARRATIVE OF EVENTS IN THE ARMY OF THE POTOMAC, FROM 1861 TO THE CLOSE OF THE REBELLION, APRIL, 1865.

by
GEORGE T. STEVENS,
SURGEON OF THE 77TH REGIMENT NEW YORK VOLUNTEERS.

1866.

MAJ.-GEN. JOHN SEDGWICK

THREE YEARS in the 6th CORPS

PREFACE.

The following pages are offered to my old comrades of the Sixth Corps, with the hope that they may pleasantly recall the many varied experiences of that unparalleled body of men. If much has been omitted which should have been written, or if anything has been said which should have been left out, I rely upon the generosity of brave men to treat with leniency the failings they may detect.

I have endeavored to present without exaggeration or embellishment of imagination, a truthful picture of army life in all its vicissitudes; its marches, its battles, its camps, and the sad scenes when the victims of war languish in hospitals. The story is written mostly from extensive notes taken by myself amid the scenes described; but official reports and letters from officers have been used freely in correcting these notes, and gathering fresh material. The narrative commences with the experiences of my own regiment; then when that regiment became a part of Smith's division, its incidents and history includes the whole. From the organization of the Sixth Corps to the close of the rebellion, I have endeavored without partiality to give the story of the Corps. If I have failed to do justice to any of the noble troops of the Corps, it has been from no want of desire to give to each regiment the praise due to it.

I cannot close without acknowledging my many obligations to the numerous friends, officers and soldiers of the Corps, and others who have favored me with their assistance. I take especial pleasure in acknowledging the kindness of Miss Emily Sedgwick, sister of our lamented commander; Vermont's honored son, Major-General L. A. Grant, Major-General Thomas H. Neill, Colonel James B. McKean, Colonel W. B. French, Chaplain Norman Fox, and Mr. Henry M. Myers. I am also indebted to the friends of Samuel S. Craig for the use of his diary, extending from the early history of the Army of the Potomac, to the death of the talented young soldier in the Wilderness.

The engravings are nearly all from sketches taken by myself on the ground, the others are from the pencil of the well known artist, Captain J. Hope, and all have been submitted to his finishing touch. Mr. Ferguson has executed the wood cuts in a style creditable to his art.

The typographical portion of the work has been done in a style of beauty and finish for which the work of Weed, Parsons and Company is so well known.

18 North Pearl Street, Albany, N. Y.
September 5, 1866.

CHAPTER XIII.
THE MARYLAND CAMPAIGN.

General McClellan restored to command—March through Washington—Leisurely campaigning—Battle of Crampton Pass— Death of Mathison—Battle of South Mountain Pass—Death of Reno— Surrender of Harper's Ferry—March to Antietam.

General Pope, at his own request, was relieved from the command of the army, and General McClellan resumed the direction. Whatever might have been the real fitness of General Pope to command, his usefulness with the army just driven back upon the defenses of Washington, had departed. The return of General McClellan was hailed with joy by a large portion of the army.

On the 5th of September, Lee crossed the Potomac into Maryland, and occupied Frederick City. General McClellan was ordered to push forward at once and meet him. It was on the evening of the 6th that orders were issued to move. It was but short work to pack up our limited supply of clothing, cooking utensils and the few other articles which constituted our store of worldly goods, and prepare to march. We left Alexandria, and proceeding toward Washington, passed Fort Albany and crossed the Long Bridge, the moon and stars shining with a brilliancy seldom equaled, rendering the night march a pleasant one. As the steady tramp of the soldiers upon the pavements was heard by the citizens of Washington, they crowded upon the walks, eager to get a glance, even by moonlight, of the veterans who had passed through such untold hardships. Many were the questions regarding our destination, but we could only answer, "We are going to meet the rebels." Passing through Georgetown, we reached the little village of Tanleytown, where, weary from the short but rapid march, we spent the remainder of the night in sleep. The morning passed without orders to move, and it was not until five o'clock in the afternoon that we again commenced the march, when, having proceeded six miles, we halted. At daybreak on the morning of the 8th, the corps was moving again, and passing through Rockville we halted, after an easy stage of six miles.

On the 9th we marched three miles, making our camp at Johnstown. On the following morning, at 9 o'clock, we were again on the move, driving before us small bodies of rebel cavalry, and reaching Barnesville, a small village, ten miles from our encampment of the night before. Our Third brigade, of the Second division, was quartered on the plantation of a noted secessionist, who, on our approach, had suddenly decamped, leaving at our disposal a very large orchard, whose trees were loaded with delicious fruit, and his poultry yard well stocked with choice fowls. Our boys were not slow to appropriate to their own use these luxuries, which, they declared, were great improvements on pork and hard tack. In the enjoyment of ease and abundance, we remained here until the morning of the 12th, when we resumed the march, proceeding ten miles farther, halting near Urbana, at Monocacy bridge, which had been destroyed by the rebels, but was now rebuilt. On the same day General Burnside, having the advance, entered Frederick, encountering a few skirmishers of the enemy, which he drove. On the 13th, we arrived at the lovely village of Jefferson, having made ten miles more, and having driven a detachment of rebels through Jefferson Pass.

The advance was sounded at ten o'clock on the morning of the 14th, and at three we found ourselves near the foot of the South Mountain range, having marched about fifty miles in eight days. Upon the advance of Burnside into Frederick, the rebel force had fallen back, taking the two roads which led through Middletown and Burkettsville, and which crossed the South Mountains through deep gorges, the northern called South Mountain or Turner's Pass, and the other, six miles south of it, Crampton Pass.

These passes the rebels had strongly fortified, and had arranged their batteries on the crests of neighboring hills. The Sixth corps came to a halt when within about a mile and a half of Crampton Pass, and a reconnoissance was ordered.

General Franklin was now directed to force the pass with the Sixth corps, while the remaining corps should push on to the South Mountain Pass and drive the enemy through it. We formed in line of battle and advanced. Before us lay the little village of Burkettsville, nestling under the shadow of those rugged mountains, its white houses gleaming out of the dark green foliage. Beyond were the South Mountains; their summits crowned with batteries of artillery and gray lines of rebels, while the heavily wooded sides concealed great numbers of the enemy.

A winding road, leading up the mountain side and through a narrow defile, known as Crampton's Gap, constituted one of the two passages to the other side of the range; South Mountain Gap being the other. The enemy had planted batteries and posted troops behind barricades, and in such positions as most effectually to dispute our passage.

At the foot of the mountain, was a stone wall, behind which was the first rebel line of battle, while their skirmishers held the ground for some distance in front. The position was a strong one; admirably calculated for defense, and could be held by a small force against a much larger one.

Charge of the Sixth Corps at Burkettsville

The day was far advanced when the attack was ordered. No sooner had the lines of blue uniforms emerged from the cover of the woods, than the batteries on the hill tops opened upon them. The mountains, like huge volcanoes, belched forth fire and smoke. The earth trembled beneath us, and the air was filled with the howling of shells which flew over our heads, and ploughed the earth at our feet. At the same time, the line of battle behind the stone wall opened upon us a fierce fire of musketry. In the face of this storm of shells and bullets, the corps pressed forward at double quick, over the ploughed grounds and through the corn fields, halting for a few moments at the village. The citizens, regardless of the shells which were crashing through their houses, welcomed us heartily, bringing water to fill the canteens, and supplying us liberally from the scanty store left them by the marauding rebels.

Patriotic ladies cheered the Union boys and brought them food; and well might they rejoice at the approach of the Union army, after their recent experience with the rebels, who had robbed them of almost everything they possessed in the way of movable property.

After a few minutes, in which our soldiers took breath, the advance was once more sounded, and again we pushed on in face of a murderous fire, at the same time pouring into the face of the foe a storm of leaden hail. Slocum's division, of the Sixth corps, advanced on the right of the turnpike, while Smith's division pushed directly forward on the road and on the left of it. After severe fighting by both divisions, having driven the enemy from point to point, Slocum's troops, about three o'clock, succeeded in seizing the pass, while our Second division pressed up the wooded sides of the mountain, charging a battery at the left of the pass and capturing two of its guns. The confederates fled precipitately down the west side of the mountain, and our flags were waved in triumph from the heights which had so lately thundered destruction upon us. As we advanced, we wondered, not that the foe had offered such stubborn resistance, but that the position had been yielded at all. Their dead strewed our path, and great care was required, as we passed along the road, to avoid treading upon the lifeless remains which lay thickly upon the ground. On every side the evidences of the fearful conflict multiplied. Trees were literally cut to pieces by shells and bullets; a continual procession of rebel wounded and prisoners lined the roadsides, while knapsacks, guns, canteens and haversacks were scattered in great confusion. The rebel force made its way into Pleasant Valley, leaving in our hands their dead and wounded, three stand of colors, two pieces of artillery and many prisoners. Our troops scoured the woods until midnight, bringing in large numbers of stragglers.

We had lost quite heavily; some of our best men had fallen. Colonel Mathison, who commanded the Third brigade of Slocum's division, whose heroism at Gaines' Farm, and bravery in all our campaign on the Peninsula, had endeared him to his division, was among the killed.

The corps moved down the road to the western side of the mountains, our men resting on their arms for the night, expecting that the battle would be renewed at dawn. But the morning revealed no enemy in our front; we were in quiet possession of the valley.

Meanwhile on the right, at South Mountain Pass, a still more sanguinary battle had been in progress.

On the morning of the 14th, the Ninth corps, Burnside's veterans, the heroes of Roanoke and Newbern, under the command of the gallant Reno, advanced from Middletown; and coming near the base of the mountains, found the enemy strongly posted on the crests of the hills, thronging the thickly wooded sides, and crowding in the gap. No matter what position the brave boys occupied, they were submitted to a murderous fire from the crests and sides of the mountains. Under this galling fire, the First division of the corps formed in line of battle, and advanced toward the frowning heights. It was an undertaking requiring more than ordinary valor, to attempt to wrest from an enemy strong in numbers, a position so formidable for defense; but the men approaching those rugged mountain sides had become accustomed to overcome obstacles, and to regard all things as possible which they were commanded to do. Under cover of a storm of shells, thrown upward to the heights, the line of battle advanced, with courage and firmness, in face of terrible resistance, gaining much ground and driving the rebels from their first line of defenses. Now, the corps of Hooker rushed to the assistance of the Ninth. As the gallant general and his staff rode along the lines, enthusiastic cheers for "Fighting Joe Hooker," greeted him everywhere. Forming his divisions hastily, he pushed them on the enemy's lines at once.

Thus far, the battle had been principally maintained by artillery; the rattle of musketry coming occasionally from one or another part of Reno's line. But now, the whole line was pushing against the rebel line, and the continued roll of musketry told of close work for the infantry. Reno's troops on the left and Hooker's on the right, were doing noble fighting. The advancing line never wavered; but pressing steadily forward, pouring volley after volley into the enemy's ranks, it at last forced the rebels to break and fly precipitately to the crests, and, leaving their splendid position on the summit to retreat in great haste down the other slope of the mountain. The engagement had been of three hours duration; and the bravery of the Union troops was rewarded by the possession of the mountain tops. Darkness put an end to the pursuit. Thus the two chief passes through the mountains were in the possession of the Union army.

While his corps was striving to dislodge the enemy from the stronghold, the gallant Reno was struck by a minie ball, and expired. The loss of this hero threw a gloom not only over his own corps, but throughout the army.

In the many battles in which he had taken a brilliant part, he had won an enviable fame, and his private virtues and kindly qualities of heart added lustre to the brilliancy of his military record.

While the fight was in progress in Crampton Pass, the booming of guns at Harper's Ferry, only seven miles distant, told us of an attempt, on the part of the rebels, to capture that important point; and while we lay upon our arms on the morning of the 15th, two miles nearer than we were on the day before, the firing was heard to be still more fierce. Our Sixth corps was ordered to press forward to the relief of the beleaguered place; but before we had started the firing suddenly died away. General Franklin concluded that the place had been surrendered; and his conclusion was verified by reconnoissances. So the corps remained in Pleasant Valley, at rest, all of the 15th and 16th.

The surrender of Harper's Ferry was a terrible blow to our cause. Had it continued in our possession it must have insured, with any respectable energy on the part of our commanders, the destruction of the rebel army in its retreat. As it was, our loss was over eleven thousand men, and a vast amount of war material.

Of course, the surrender of Harper's Ferry, at this critical period, was owing directly to the imbecility and cowardice, not to say treachery, of the officers in command at Harper's Ferry and on Maryland Heights. But, while we condemn the weakness and cowardice of these commanders, can we relieve from a share in the responsibility, the general who marched his army in pursuit of the enemy at a snail pace, traveling but six miles a day upon an average, when by a few brisk marches this important point might have been reinforced?

Early on the morning of the 17th, the Sixth corps was on its way, hastening to the scene of conflict which had commenced on the banks of Antietam creek. A part of the Seventy-seventh had constituted one-third of the picket line which had extended across the valley between the corps and Harper's Ferry.

These companies, by a hard march, much of it at double quick, succeeded in overtaking the division just as the Third brigade was making a charge over ground already thrice won and lost by Sumner's troops. Without waiting to form the companies, the detachment joined the command, and, all out of breath and faint from their forced march, rushed with their companies against the foe.

This Account Taken from:

A LIFE OF GEN. ROBERT E. LEE.

by
JOHN ESTEN COOKE

"Duty is the sublimest word in our language."
"Human virtue should be equal to human calamity."
LEE.

1876

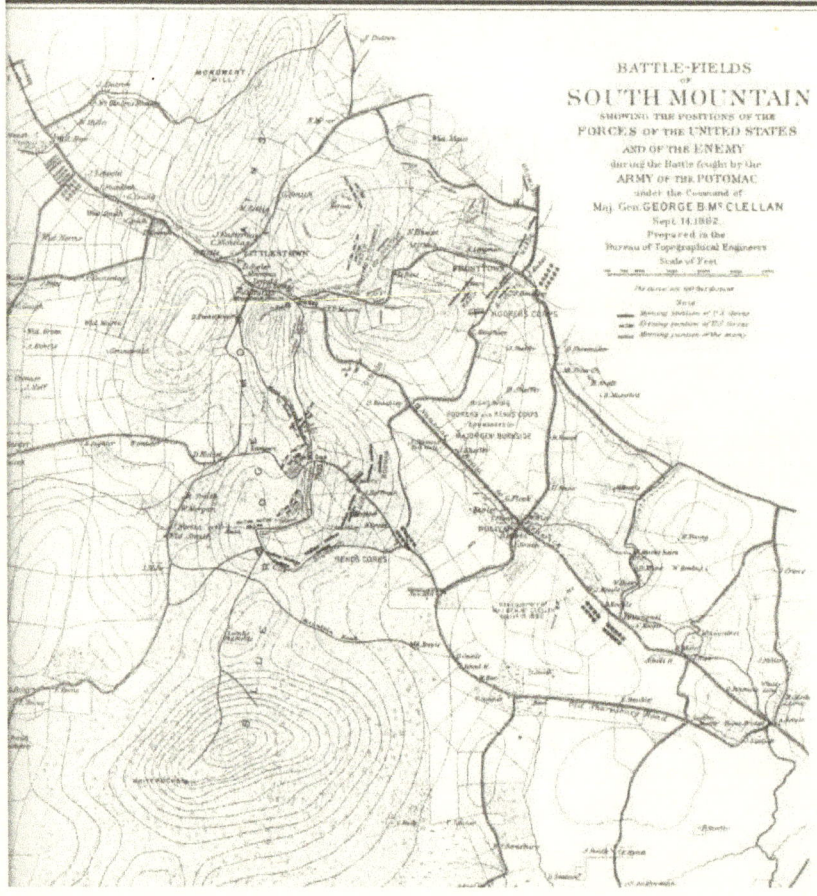

BATTLE-FIELDS
OF
SOUTH MOUNTAIN
SHOWING THE POSITIONS OF THE
FORCES OF THE UNITED STATES
AND OF THE ENEMY
during the Battle fought by the
ARMY OF THE POTOMAC
under the Command of
Maj. Gen. GEORGE B. McCLELLAN
Sept. 14, 1862.
Prepared in the
Bureau of Topographical Engineers
Scale of Feet

PART V.
LEE INVADES MARYLAND.
I
HIS DESIGNS.

The defeat of General Pope opened the way for movements not contemplated, probably, by General Lee, when he marched from Richmond to check the advance in Culpepper. His object at that time was doubtless simply to arrest the forward movement of the new force threatening Gordonsville. Now, however, the position of the pieces on the great chess-board of war had suddenly changed, and it was obviously Lee's policy to extract all the advantage possible from the new condition of things.

He accordingly determined to advance into Maryland—the fortifications in front of Washington, and the interposition of the Potomac, a broad stream easily defended, rendering a movement in that direction unpromising. On the 3d of September, therefore, and without waiting to rest his army, which was greatly fatigued with the nearly continuous marching and fighting since it had left the Rapidan, General Lee moved toward Leesburg, crossed his forces near that place, and to the music of the bands playing the popular air, "Maryland, my Maryland," advanced to Frederick City, which he occupied on the 7th of September.

Lee's object in invading Maryland has been the subject of much discussion, one party holding the view that his sole aim was to surround and capture a force of nine or ten thousand Federal troops stationed at Harper's Ferry; and another party maintaining that he proposed an invasion of Pennsylvania as far as the Susquehanna, intending to fight a decisive battle there, and advance thereafter upon Philadelphia, Baltimore, or Washington. The course pursued by an army commander is largely shaped by the progress of events. It can only be said that General Lee, doubtless, left the future to decide his ultimate movements;

meanwhile he had a distinct and clearly-defined aim, which he states in plain words.

His object was to draw the Federal forces out of Virginia first. The movement culminating in the victory over the enemy at Manassas had produced the effect of paralyzing them in every quarter. On the coast of North Carolina, in Western Virginia, and in the Shenandoah Valley, had been heard the echo of the great events in Middle and Northern Virginia. General Burnside's force had been brought up from the South, leaving affairs at a stand-still in that direction; and, contemporaneously with the retreat of General Pope, the Federal forces at Washington and beyond had fallen back to the Potomac. This left the way open, and Lee's farther advance, it was obvious, would now completely clear Virginia of her invaders. The situation of affairs, and the expected results, are clearly stated by General Lee:

"The war was thus transferred," he says, "from the interior to the frontier, and the supplies of rich and productive districts made accessible to our army. To prolong a state of affairs in every way desirable, and not to permit the season for active operations to pass without endeavoring to inflict other injury upon the enemy, the best course appeared to be the transfer of the army into Maryland."

The state of things in Maryland was another important consideration. That great Commonwealth was known to be sectionally divided in its sentiment toward the Federal Government, the eastern portion adhering generally to the side of the South, and the western portion generally to the Federal side. But, even as high up as Frederick, it was hoped that the Southern cause would find adherents and volunteers to march under the Confederate banner. If this portion of the population had only the opportunity to choose their part, unterrified by Federal bayonets, it was supposed they would decide for the South. In any event, the movement would be important. The condition of affairs in Maryland, General Lee says, "encouraged the belief that the presence of our army, however inferior to that of the enemy, would induce the Washington Government to retain all its available force to provide for contingencies which its course toward the people of that State gave it reason to apprehend," and to cross the Potomac "might afford us an opportunity to aid the citizens of Maryland in any efforts they might be disposed to make to recover their liberty."

It may be said, in summing up on this point, that Lee expected volunteers to enroll themselves under his standard, tempted to do so by the hope of throwing off the yoke of the Federal Government, and the army certainly shared this expectation. The identity of sentiment generally between the people of the States of Maryland and Virginia, and their strong social ties in the past, rendered this anticipation reasonable, and the feeling of the country at the result afterward was extremely bitter.

Such were the first designs of Lee; his ultimate aim seems as clear. By advancing into Maryland and threatening Baltimore and Washington, he knew that he would force the enemy to withdraw all their troops from the south bank of the Potomac, where they menaced the Confederate communications with Richmond; when this was accomplished, as it clearly would be, his design was, to cross the Maryland extension of the Blue Ridge, called there the South Mountain, advance by way of Hagerstown into the Cumberland Valley, and, by thus forcing the enemy to follow him, draw them to a distance from their base of supplies, while his own communications would remain open by way of the Shenandoah Valley. This was essentially the same plan pursued in the campaign of 1863, which terminated in the battle of Gettysburg. General Lee's movements now indicated similar intentions. He doubtless wished, in the first place, to compel the enemy to pursue him—then to lead them as far as was prudent—and then, if circumstances were favorable, bring them to decisive battle, success in which promised to open for him the gates of Washington or Baltimore, and end the war.

It will now be seen how the delay caused by the movement of Jackson against Harper's Ferry, and the discovery by General McClellan of the entire arrangement devised by Lee for that purpose, caused the failure of this whole ulterior design.

II.
LEE IN MARYLAND.

The Southern army was concentrated in the neighborhood of Frederick City by the 7th of September, and on the next day General Lee issued an address to the people of Maryland.

We have not burdened the present narrative with Lee's army orders and other official papers; but the great force and dignity of this address render it desirable to present it in full:

HEADQUARTERS ARMY OF NORTHERN VIRGINIA,}
NEAR FREDERICKTOWN, September 8, 1862.}

> *To the People of Maryland:*
>
> *It is right that you should know the purpose that has brought the army under my command within the limits of your State, so far as that purpose concerns yourselves.*
>
> *The people of the Confederate States have long watched with the deepest sympathy the wrongs and outrages that have been inflicted upon the citizens of a Commonwealth allied to the States of the South by the strongest social, political, and commercial ties.*
>
> *They have seen, with profound indignation, their sister State deprived of every right, and reduced to the condition of a conquered province. Under the pretence of supporting the Constitution, but in violation of its most valuable provisions, your citizens have been arrested and imprisoned upon no charge, and contrary to all forms of law. The faithful and manly protest against this outrage, made by the*

71

venerable and illustrious Marylanders—to whom in better days no citizen appealed for right in vain—was treated with scorn and contempt. The government of your chief city has been usurped by armed strangers; your Legislature has been dissolved by the unlawful arrest of its members; freedom of the press and of speech have been suppressed; words have been declared offences by an arbitrary desire of the Federal Executive, and citizens ordered to be tried by military commission for what they may dare to speak.

Believing that the people of Maryland possessed a spirit too lofty to submit to such a government, the people of the South have long wished to aid you in throwing off this foreign yoke, to enable you again to enjoy the inalienable rights of freemen, and restore independence and sovereignty to your State.

In obedience to this wish, our army has come among you, and is prepared to assist you, with the power of its arms, in regaining the rights of which you have been despoiled. This, citizens of Maryland, is our mission, so far as you are concerned. No constraint upon your free will is intended—no intimidation will be allowed. Within the limits of this army, at least, Marylanders shall once more enjoy their ancient freedom of thought and speech. We know no enemies among you, and will protect all of every opinion. It is for you to decide your destiny, freely, and without constraint. This army will respect your choice, whatever it may be; and, while the Southern people will rejoice to welcome you to your natural position among them, they will only welcome you when you come of your own free will.

R.E. LEE, General commanding.

This address, full of grave dignity, and highly characteristic of the Confederate commander, was in vivid contrast with the harsh orders of

General Pope in Culpepper. The accents of friendship and persuasion were substituted for the "rod of iron." There would be no coercive measures; no arrests, with the alternative presented of an oath to support the South, or instant banishment. No intimidation would be permitted. In the lines of the Southern army, at least, Marylanders should enjoy freedom of thought and speech, and every man should "decide his destiny freely, and without constraint."

This address, couched in terms of such dignity, had little effect upon the people. Either their sentiment in favor of the Union was too strong, or they found nothing in the condition of affairs to encourage their Southern feelings. A large Federal force was known to be advancing; Lee's army, in tatters, and almost without supplies, presented a very uninviting appearance to recruits, and few joined his standard, the population in general remaining hostile or neutral.

The condition of the army was indeed forlorn. It was worn down by marching and fighting; the men had scarcely shoes upon their feet; and, above the tattered figures, flaunting their rags in the sunshine, were seen gaunt and begrimed faces, in which could be read little of the "romance of war." The army was in no condition to undertake an invasion; "lacking much of the material of war, feeble in transportation, poorly provided with clothing, and thousands of them destitute of shoes," is Lee's description of his troops. Such was the condition of the better portion of the force; on the opposite side of the Potomac, scattered along the hills, could be seen a weary, ragged, hungry, and confused multitude, who had dragged along in rear of the rest, unable to keep up, and whose miserable appearance said little for the prospects of the army to which they belonged.

From these and other causes resulted the general apathy of the Marylanders, and Lee soon discovered that he must look solely to his own men for success in his future movements. He faced that conviction courageously; and, without uttering a word of comment, or indulging in any species of crimination against the people of Maryland, resolutely commenced his movements looking to the capture of Harper's Ferry and the invasion of Pennsylvania.[1]

[Footnote 1: The reader will perceive that the intent to invade Pennsylvania is repeatedly attributed in these pages to General Lee. His own expression is, "by threatening Pennsylvania, to induce the enemy," etc. That he designed invasion, aided by the recruits anticipated in Maryland, seems unquestionable; since, even after discovering the lukewarmness of the people there by the fact that few joined his standard, he still advanced to Hagerstown, but a step from the Pennsylvania line. These facts have induced the present writer to attribute the design of actual invasion to Lee with entire confidence; and all the circumstances seem to him to support that hypothesis.]

The promises of his address had been kept. No one had been forced to follow the Southern flag; and now, when the people turned their backs upon it, closing the doors of the houses in the faces of the Southern troops, they remained unmolested. Lee had thus given a practical proof of the sincerity of his character. He had promised nothing which he had not performed; and in Maryland, as afterward in Pennsylvania, in 1863, he remained firm against the temptation to adopt the harsh course generally pursued by the commanders of invading armies. He seems to have proceeded on the principle that good faith is as essential in public affairs as in private, and to have resolved that, in any event, whether of victory or disaster, his enemies should not have it in their power to say that he broke his plighted word, or acted in a manner unbecoming a Christian gentleman.

Prompt action was now necessary. The remnants of General Pope's army, greatly scattered and disorganized by the severe battle of Manassas, had been rapidly reformed and brought into order again, and to this force was added a large number of new troops, hurried forward from the Northern States to Washington. This new army was not to be commanded by General Pope, who had been weighed and found wanting in ability to contend with Lee. The force was intrusted to General McClellan, in spite of his unpopularity with the Federal authorities; and the urgent manner in which he had been called upon to take the head of affairs and protect the Federal capital, is the most eloquent of all commentaries upon the position which he held in the eyes of the country and the army. It was felt, indeed, by all that the Federal ship was rolling in the storm, and an experienced pilot was necessary for her guidance. General McClellan was accordingly directed, after General Pope's defeat, to take command of every thing, and see to the safety of Washington; and, finding himself at length at the head of an army of about one hundred thousand men, he proceeded, after the manner of a good soldier, to protect the Federal capital by advancing into upper Maryland in pursuit of Lee.

III.
MOVEMENTS OF THE TWO ARMIES.

General Lee was already moving to the accomplishment of his designs, the capture of Harper's Ferry, and an advance into the Cumberland Valley.

His plan to attain the first-mentioned object was simple, and promised to be successful. Jackson was to march around by way of "Williamsport and Martinsburg," and thus approach from the south. A force was meanwhile to seize upon and occupy the Maryland Heights, a lofty spot of the mountain across the Potomac, north of the Ferry. In like manner, another body of troops was to cross the Potomac, east of the Blue Ridge, and occupy the Loudon Heights, looking down upon Harper's Ferry from the east. By this arrangement the retreat of the enemy would be completely cut off in every direction. Harper's Ferry must be captured, and, having effected that result, the whole Confederate force, detached for the purpose, was to follow the main body of this army in the direction of Hagerstown, to take part in the proposed invasion of Pennsylvania.

This excellent plan failed, as will be seen, from no fault of the great soldier who devised it, but in consequence of unforeseen obstacles, and especially of one of those singular incidents which occasionally reverse the best-laid schemes and abruptly turn aside the currents of history.

Jackson and the commanders coöperating with him moved on September 10th. General Lee then with his main body crossed the South Mountain, taking the direction of Hagerstown. Meanwhile, General McClellan had advanced cautiously and slowly, withheld by incessant dispatches from Washington, warning him not to move in such a manner as to expose that city to danger. Such danger existed only in the imaginations of the authorities, as the army in advancing extended its front from the Potomac to the Baltimore and Ohio Railroad. General McClellan, nevertheless, moved with very great precaution, feeling his way, step by step, like a man in the dark, when on reaching Frederick City, which the Confederates had just evacuated, good fortune suddenly came to his assistance. This good fortune was the discovery of a copy of General Lee's orders of march for the army, in which his whole plan was revealed. General McClellan had therein the unmistakable evidence of his opponent's intentions, and from that moment his advance was as rapid as before it had been deliberate.

The result of this fortunate discovery was speedily seen. General Lee, while moving steadily toward Hagerstown, was suddenly compelled to turn his attention to the mountain-passes in his rear. It had not been the intention of Lee to oppose the passage of the enemy through the South Mountain, as he desired to draw General McClellan as far as possible from his base, but the delay in the fall of Harper's Ferry now made this necessary. It was essential to defend the mountain-defiles in order to insure the safety of the Confederate troops at Harper's Ferry; and Lee accordingly directed General D.H. Hill to oppose the passage of the enemy at Boonsboro Gap, and Longstreet was sent from Hagerstown to support him.

An obstinate struggle now ensued for the possession of the main South Mountain Gap, near Boonsboro, and the roar of Jackson's artillery from Harper's Ferry must have prompted the assailants to determined efforts to force the passage. The battle continued until night (September 14th), and resulted in heavy loss on both sides, the brave General Reno, of the United States army, among others, losing his life. Darkness put an end to the action, the Federal forces not having succeeded in passing the Gap; but, learning that a column of the enemy had crossed below and threatened him with an attack in flank, General Lee determined to retire in the direction of Sharpsburg, where Jackson and the forces coöperating with him could join the main body of the army. This movement was effected without difficulty, and Lee notices the skill and efficiency of General Fitz Lee in covering the rear with his cavalry. The Federal army failed to press forward as rapidly as it is now obvious it should have done. The head of the column did not appear west of the mountain until eight o'clock in the morning (September 15th), and, nearly at the same moment ("the attack began at dawn; in about two hours the garrison surrendered," says General Lee), Harper's Ferry yielded to Jackson.

Fast-riding couriers brought the welcome intelligence of Jackson's success to General Lee, as the latter was approaching Sharpsburg, and official information speedily came that the result had been the capture of more than eleven thousand men, thirteen thousand small-arms, and seventy-three cannon. It was probably this large number of men and amount of military stores falling into the hands of the Confederates which afterward induced the opinion that Lee's sole design in invading Maryland had been the reduction of Harper's Ferry.

General McClellan had thus failed, in spite of every effort which he had made, to relieve Harper's Ferry,[1] and no other course remained now but to follow Lee and bring him to battle. The Federal army accordingly moved on the track of its adversary, and, on the afternoon of the same day (September 15th), found itself in sight of Lee's forces drawn up on the western side of Antietam Creek, near the village of Sharpsburg.

[Footnote 1: All along the march he had fired signal-guns to inform the officer in command at Harper's Ferry of his approach.]

At last the great opponents were in face of each other, and a battle, it was obvious, could not long be delayed.

General McClellen Rides through the City of Frederick, Md.

This Account is Taken from:

HISTORY OF KERSHAW'S BRIGADE, WITH COMPLETE ROLL OF COMPANIES, BIOGRAPHICAL SKETCHES, INCIDENTS, ANECDOTES, ETC.

by
D. AUGUSTUS DICKERT

General George McClellan

CHAPTER X

The March to Maryland—Second Manassas. Capture of Harper's Ferry—Sharpsburg

On the 5th or 6th we rejoined at last, after a two months' separation from the other portion of the army. Lee was now preparing to invade Maryland and other States North, as the course of events dictated. Pope's Army had joined that of McClellan, and the authorities at Washington had to call on the latter to "save their Capital." When the troops began the crossing of the now classic Potomac, a name on every tongue since the commencement of hostilities, their enthusiasm knew no bounds. Bands played "Maryland, My Maryland," men sang and cheered, hats filled the air, flags waved, and shouts from fifty thousand throats reverberated up and down the banks of the river, to be echoed back from the mountains and die away among the hills and highlands of Maryland. Men stopped midway in the stream and sang loudly the cheering strains of Randall's, "Maryland, My Maryland." We were overjoyed at rejoining the army, and the troops of Jackson, Longstreet, and the two Hills were proud to feel the elbow touch of such chivalrous spirits as McLaws, Kershaw, Hampton, and others in the conflicts that were soon to take place. Never before had an occurrence so excited and enlivened the spirits of the troops as the crossing of the Potomac into the land of our sister, Maryland. It is said the Crusaders, after months of toil, marching, and fighting, on their way through the plains of Asia Minor, wept when they saw the towering spires of Jerusalem, the Holy City, in the distance; and if ever Lee's troops could have wept for joy, it was at the crossing of the Potomac. But we paid dearly for this pleasure in the death of so many thousands of brave men and the loss of so many valuable officers. General Winder fell at Cedar Mountain, and Jackson's right hand, the brave Ewell, lost his leg at Manassas.

The army went into camp around Frederick City, Md. From here, on the 8th, Lee issued his celebrated address to the people of Maryland, and to those of the North generally, telling them of his entry into their country, its cause and purpose; that it was not as a conqueror, or an enemy, but to demand and enforce a peace between the two countries. He clothed his language in the most conservative and entreating terms, professing friendship for those who would assist him, and protection to life and the property of all. He enjoined the people, without regard to past differences, to flock to his standard and aid in the defeat of the party and people who were now drenching the country in blood and putting in mourning the people of two nations. The young men he asked to join his ranks as soldiers of a just and honorable cause. Of the old he asked their sympathies and prayers. To the President of the Confederate States he also wrote a letter, proposing to him that he should head his armies, and, as the chieftain of the nation, propose a peace to the authorities at Washington from the very threshold of their Capital. But both failed of the desired effect. The people of the South had been led to believe that Maryland was anxious to cast her destinies with those of her sister States, that all her sympathies were with the people of the South, and that her young men were anxious and only awaiting the opportunity to join the ranks as soldiers under Lee. But these ideas and promises were all delusions, for the people we saw along the route remained passive spectators and disinterested witnesses to the great evolutions now taking place. What the people felt on the "eastern shore" is not known; but the acts of those between the Potomac and Pennsylvania above Washington indicated but little sympathy with the Southern cause; and what enlistments were made lacked the proportions needed to swell Lee's army to its desired limits. Lee promised protection and he gave it. The soldiers to a man seemed to feel the importance of obeying the orders to respect and protect the person and property of those with whom we came in contact. It was said of this, as well as other campaigns in the North, that "it was conducted with kid gloves on."

While lying at Frederick City, Lee conceived the bold and perilous project of again dividing his army in the face of his enemy, and that enemy McClellan. Swinging back with a part of his army, he captured the stronghold of Harper's Ferry, with its 11,000 defenders, while with the other he held McClellan at bay in front. The undertaking was dangerous in the extreme, and with a leader less bold and Lieutenants less prompt and skillful, its final consummation would have been more than problematical. But Lee was the one to propose his subalterns to act. Harper's Ferry, on the Virginia side of the Potomac, where that river is intersected by the Shenandoah, both cutting their way through the cliffs and crags of the Blue Ridge, was the seat of the United States Arsenal, and had immense stores of arms and ammunition, as well as army supplies of every description. The Baltimore and Ohio Railroad and the canal cross the mountains here on the Maryland side, both hugging the precipitous side of the mountain and at the very edge of the water. The approaches to the place were few, and they so defended that capture seemed impossible, unless the heights surrounding could be obtained, and this appeared impossible from a military point of view. On the south side are the Loudon and Bolivar Heights. On the other side the mountains divide into two distinct ranges and gradually bear away from each other until they reach a distance of three miles from crest to crest. Between the two mountains is the beautiful and picturesque Pleasant Valley. The eastern ridge, called South Mountain, commencing from the rugged cliff at Rivertoria, a little hamlet nestled down between the mountains and the Potomac, runs northwards, while the western ridge, called Elk Mountain, starts from the bluff called Maryland Heights, overlooking the town of Harper's Ferry, and runs nearly parallel to the other. Jackson passed on up the river with his division, Ewell's, and A.P. Hill's, recrossed the Potomac into Virginia, captured Martinsburg, where a number of prisoners and great supplies were taken, and came up and took possession of Bolivar Heights, above Harper's Ferry. Walker's Division marched back across the Potomac and took possession of Loudon Heights, a neck of high land between the Shenandoah and Potomac overlooking Harper's Ferry from below, the Shenandoah being between his army and the latter place. On the 11th McLaws moved out of Frederick City, strengthened by the brigades of Wilcox, Featherstone, and Pryor, making seven brigades that were to undertake the capture of the stronghold by the mountain passes and ridges on the north. Kershaw, it will be seen, was given the most difficult position of passage and more formidable to attack than any of the other routes of approach. Some time after Jackson and Walker had left on their long march, McLaws followed. Longstreet and other portions of the army and wagon trains kept the straight road towards Hagerstown, while Kershaw and the rest of the troops under McLaws took the road leading southwest, on through the town of Burkettville, and camped at the foothills of the mountain, on the east side. Next morning Kershaw, commanding his own brigade and that of Barksdale, took the lead, passed over South Mountain, through Pleasant Valley, and to Elk Ridge, three miles distance, thence along the top of Elk Ridge by a dull cattle path. The width of the crest was not more than fifty yards in places, and along this Kershaw had to move in line of battle, Barksdale's Brigade in reserve. Wright's Brigade moved along a similar path on the crest of South Mountain, he taking with him two mountain howitzers, drawn by one horse each. McLaws, as Commander-in-Chief, with some of the other brigades, marched by the road at the base of the mountain below Wright, while Cobb was to keep abreast of Kershaw and Barksdale at the base of Elk Ridge. Over such obstacles as were encountered and the difficulties and dangers separating the different troops, a line of battle never before made headway as did those of Kershaw and the troops under McLaws.

We met the enemy's skirmishers soon after turning to the left on Elk Ridge, and all along the whole distance of five miles we were more or less harassed by them. During the march of the 12th the men had to pull themselves up precipitous inclines by the twigs and undergrowth that lined the mountain side, or hold themselves in position by the trees in front. At night we bivouaced on the mountain. We could see the fires all along the mountain side and gorges through Pleasant Valley and up on South Mountain, where the troops of Wright had camped opposite. Early next morning as we advanced we again met the enemy's skirmishers, and had to be continually driving them back. Away to the south and beyond the Potomac we could hear the sound of Jackson's guns as he was beating his way up to meet us. By noon we encountered the enemy's breastworks, built of great stones and logs, in front of which was an abattis of felled timber and brushwood. The Third, under Nance, and the Seventh, under Aiken, were ordered to the charge on the right. Having no artillery up, it was with great difficulty we approached the fortifications. Men had to cling to bushes while they loaded and fired. But with their usual gallantry they came down to their work. Through the tangled undergrowth, through the abattis, and over the breastworks they leaped with a yell. The fighting was short but very severe. The Third did not lose any field officers, but the line suffered considerably. The Third lost some of her most promising officers. Of the Seventh, Captain Litchfield, of Company L, Captain Wm. Clark, of Company G, and lieutenant J.L. Talbert fell dead, and many others wounded.

The Second and Eighth had climbed the mountains, and advanced on Harper's Ferry from the east. The Second was commanded by Colonel Kennedy and the Eighth by Colonel Henagan. The enemy was posted behind works, constructed the same as those assaulted by the Third and Seventh, of cliffs of rocks, trunks of trees, covered by an abattis. The regiments advanced in splendid style, and through the tangled underbrush and over boulders they rushed for the enemy's works. Colonel Kennedy was wounded in the early part of the engagement, but did not leave the field. The Second lost some gallant line officers. When the order was given to charge the color bearer of the Eighth, Sergeant Strother, of Chesterfield, a tall, handsome man of six feet three in height, carrying the beautiful banner presented to the regiment by the ladies of Pee Dee, fell dead within thirty yards of the enemy's works. All the color guard were either killed or wounded. Captain A.T. Harllee, commanding one of the color companies, seeing the flag fall, seized it and waving it aloft, called to the men to forward and take the breastworks. He, too, fell desperately wounded, shot through both thighs with a minnie ball. He then called to Colonel Henagan, he being near at hand, to take the colors. Snatching them from under Captain Harllee, Colonel Henagan shouted to the men to follow him, but had not gone far before he fell dangerously wounded. Some of the men lifted up their fallen Colonel and started to the rear; but just at this moment his regiment began to waver and break to the rear. The gallant Colonel seeing this ordered his men to put him down, and commanded in a loud, clear voice, "About face! Charge and take the works," which order was obeyed with promptness, and soon the flags of Kershaw's Regiments waved in triumph over the enemy's deserted works.

Walker had occupied Loudon Heights, on the Virginia side, and all were waiting now for Jackson to finish the work assigned to him and to occupy Bolivar Heights, thus finishing the cordon around the luckless garrison. The enemy's cavalry under the cover of the darkness crossed the river, hugged its banks close, and escaped. During the night a road was cut to the top of Maryland Heights by our engineer corps and several pieces of small cannon drawn up, mostly by hand, and placed in such position as to sweep the garrison below. Some of Jackson's troops early in the night began climbing around the steep cliffs that overlook the Shenandoah, and by daylight took possession of the heights opposite to those occupied by Walker's Division. But all during the day, while we were awaiting the signal of Jackson's approach, we heard continually the deep, dull sound of cannonading in our rear. Peal after peal from heavy guns that fairly shook the mountain side told too plainly a desperate struggle was going on in the passes that protected our rear. General McLaws, taking Cobb's Georgia Brigade and some cavalry, hurried back over the rugged by-paths that had been just traversed, to find D.H. Hill and Longstreet in a hand-to-hand combat, defending the routes on South Mountain that led down on us by the mountain crests. The next day orders for storming the works by the troops beyond the river were given. McLaws and Walker had secured their position, and now were in readiness to assist Jackson. All the batteries were opened on Bolivar Heights, and from the three sides the artillery duel raged furiously for a time, while Jackson's infantry was pushed to the front and captured the works there. Soon thereafter the white flag was waving over Harper's Ferry, "the citadel had fallen." In the capitulation eleven thousand prisoners, seventy-two pieces of artillery, twelve thousand stands of small arms, horses, wagons, munitions, and supplies in abundance passed into the hands of the Confederates. Jackson's troops fairly swam in the delicacies, provisions, and "drinkables" constituting a part of the spoils taken, while Kershaw's and all of McLaw's and Walker's troops, who had done the hardest of the fighting, got none. Our men complained bitterly of this seeming injustice. It took all day to finish the capitulation, paroling prisoners, and dividing out the supplies; but we had but little time to rest, for Lee's Army was now in a critical condition. McClellan, having by accident captured Lee's orders specifying the routes to be taken by all the troops after the fall of Harper's Ferry, knew exactly where and when to strike. The Southern Army was at this time woefully divided, a part being between the Potomac and the Shenandoah, Jackson with three divisions across the Potomac in Virginia, McLaws with his own and a part of Anderson's Division on the heights of Maryland, with the enemy five miles in his rear at Crompton Pass cutting him off from retreat in that direction. Lee, with the rest of his army and reserve trains, was near Hagerstown.

On the 16th we descended the mountain, crossed the Potomac, fell in the rear of Jackson's moving army, and marched up the Potomac some distance, recrossed into Maryland, on our hunt for Lee and his army. The sun poured down its blistering rays with intense fierceness upon the already fatigued and fagged soldiers, while the dust along the pikes, that wound over and around the numerous hills, was almost stifling. We bivouaced for the night on the roadside, ten miles from Antietam Creek, where Lee was at the time concentrating his army, and where on the next day was to be fought the most stubbornly contested and bloody battle of modern times, if we take in consideration the number of troops engaged, its duration, and its casualties. After three days of incessant marching and fighting over mountain heights, rugged gorges, wading rivers—all on the shortest of rations, many of the men were content to fall upon the bare ground and snatch a few moments of rest without the time and trouble of a supper.

Stonewall Jackson

General D.H. Hill

Poem taken from:

POEMS TEACHERS ASK FOR
BOOK TWO

Selected by READERS OF "NORMAL INSTRUCTOR-PRIMARY PLANS"

CONTAINING MORE THAN TWO HUNDRED POEMS REQUESTED FOR PUBLICATION IN THAT MAGAZINE ON THE PAGE "POEMS OUR READERS HAVE ASKED FOR"

1925

General Reno

The Pride of Battery B

South Mountain towered upon our right, far off the river lay,
And over on the wooded height we held their lines at bay.
At last the muttering guns were still; the day died slow and wan;
At last the gunners pipes did fill, the sergeant's yarns began.
When, as the wind a moment blew aside the fragrant flood
Our brierwoods raised, within our view a little maiden stood.
A tiny tot of six or seven, from fireside fresh she seemed,
(Of such a little one in heaven one soldier often dreamed.)
And as we stared, her little hand went to her curly head
In grave salute. "And who are you?" at length the sergeant said.
"And where's your home?" he growled again. She lisped out, "Who is me?
Why, don't you know? I'm little Jane, the Pride of Battery B.
My home? Why, that was burned away, and pa and ma are dead;
And so I ride the guns all day along with Sergeant Ned.
And I've a drum that's not a toy, a cap with feathers, too;
And I march beside the drummer boy on Sundays at review.
But now our 'bacca's all give out, the men can't have their smoke,
And so they're cross—why, even Ned won't play with me and joke.
And the big colonel said to-day—I hate to hear him swear—
He'd give a leg for a good pipe like the Yanks had over there.
And so I thought when beat the drum, and the big guns were still,
I'd creep beneath the tent and come out here across the hill
And beg, good Mister Yankee men, you'd give me some 'Lone Jack.'
Please do: when we get some again, I'll surely bring it back.
Indeed I will, for Ned—says he,—if I do what I say,
I'll be a general yet, maybe, and ride a prancing bay."

We brimmed her tiny apron o'er; you should have heard her laugh
As each man from his scanty store shook out a generous half.
To kiss the little mouth stooped down a score of grimy men,
Until the sergeant's husky voice said,"'Tention squad!" and then

We gave her escort, till good-night the pretty waif we bid,
And watched her toddle out of sight—or else 'twas tears that hid
Her tiny form—nor turned about a man, nor spoke a word,
Till after awhile a far, hoarse shout upon the wind we heard!
We sent it back, then cast sad eyes upon the scene around;
A baby's hand had touched the ties that brothers once had bound.

That's all—save when the dawn awoke again the work of hell,
And through the sullen clouds of smoke the screaming missiles fell,
Our general often rubbed his glass, and marveled much to see
Not a single shell that whole day fell in the camp of Battery B.

Frank H. Gassaway.

General Ambrose Burnside

Lt. General John B. Hood

Darius Couch

Edwin Sumner

Joseph Hooker

Maj. General J.D. Cox

Robert Rhodes

William B. Franklin

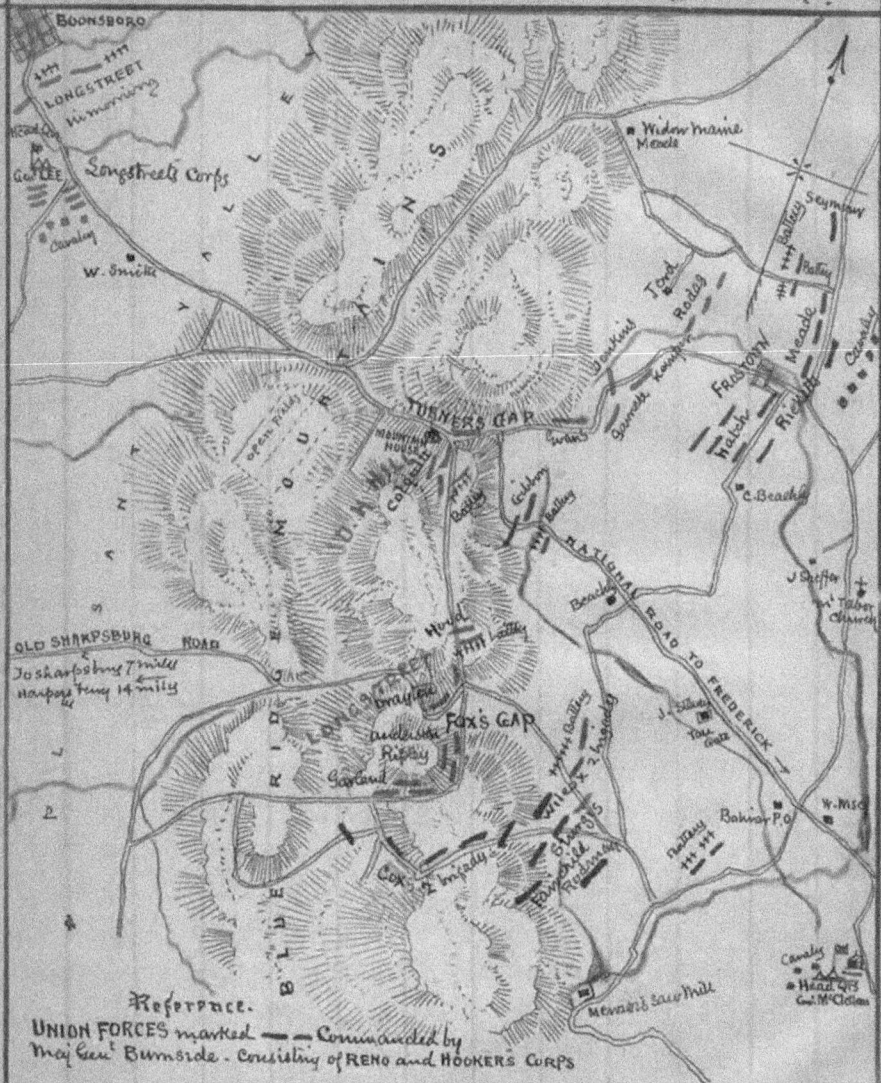

MAP OF THE POSITION AT TURNER'S GAP SOUTH MOUNTAIN

BOONSBORO

LONGSTREET Maneuvering

HEAD QRS

Gen LEE

Longstreet's Corps

Chumley

W. Smith

Widow Maine Meade

Battery Seymour

Ratey

Meade

Tack

Ridge

Robey

Hawkins

Campbell

FREEDOM

Ralph

Rumsell Kingman

TURNERS GAP

MOUNTAIN HOUSE

Colquitt

Evans

Gibbon

Battery

Guilam

C. Beachly

RAILROAD ROAD TO FREDERICK

Beachly

J. Shaffer

M. Tabor Church

OLD SHARPSBURG ROAD

To Sharpsburg 7 miles

Harpers Ferry 14 miles

Hood

Battery

Drayton

Anderson

Ripley

Garland

FOX'S GAP

Battery

Reno killed

Battery

Bolivar P.O.

W. Mse

Menard's Saw Mill

Cox 2 brigades

Cavalry

Head Qrs

Genl McClellan

Reference.

UNION FORCES marked ▬▬ Commanded by
Maj Genl Burnside. Consisting of RENO and HOOKERS CORPS

REBEL FORCES ▬▬ D. H. Hills Division and part of LONGSTREETS CORPS.

Houses ▪   Batteries ▦ ▦

SCALE OF FEET   1000   2000   3000   4000

September 14th 1862

SOUTH MOUNTAIN

Showing the Positions of the Forces of the UNITED STATES
and of the ENEMY
during the Battle fought by the Army of the Potomac
under the Command of
MAJOR GENERAL G. B. McCLELLAN
Sep. 14th 1862
Prepared in the
Bureau of Topographical Engineers

RIGHT WING
HOOKER'S and RENO'S CORPS
commanded by
MAJOR GEN. BURNSIDE

Left Wing
6th Corps
commanded by
MAJ. GEN. W. B. FRANKLIN

1862	McClellan, George B.
Sept. 15 rec'd	H.Q. Bolivar
	To Henry W. Halleck Washington, D.C.
	Tel. 1 p. 8°

Telegram from McClellen to Washington regarding the aftermath of the Battle of South Mountain.

Head Qrs Army Potomac
Poler or Sept 15ᵗʰ 8 AM

Maj Gn H W Halleck
Gen in Chief —

1862

I have just learned
from Genl Hooker in the advance,
who states that information is perfectly
reliable that the enemy is making
for Shepardstown in a perfect panic
& Genl Lee last night stated publicly
that he must admit they had been
shockingly whipped — I am hurrying
everything forward to endeavor to
press their retreat to the utmost

G B McClellan
Maj Genl

M'CLELLAN AND VICTORY!!

OR THE BATTLE OF SOUTH MOUNTAIN AND THE UPRISING OF THE KEYSTONE STATE.

TUNE—"DAN TUCKER."

I'll sing a song, how brave McCLELLAN
Gave the rebel foe a drilling,
And made them beat retreat in fear,
With this tune ringing in their ear,
 Clear the way "Little Mac's" advancing,
 To set your Stonewall Jackson dancing.'

The rebels driven to theft and plunder,
Thought to scare us by their thunder,
They crossed the Potomac to Hagerstown,
To scare the women up and down.
 Clear the way, "Little Mac's" coming, &c

They plundered barns and fields of corn,
Of Maryland whiskey, took a horn,
They seized each union patriot there.
But Freedom's Eagle sung in the air.
 Oh! clear the way, for "Mac's" advancing.

Stonewall Jackson and "Brag Lee,"
Now thought their course was clear and free,
To take Pennsylvania right off hand,
And carry Philadelphia to Dixey's Land.
 Clear the way "Little Mac" is coming.

So on they marched for their grand attack,
But the mighty mind of "Little Mac,"
Their plans and dodges kept his eye on,
As still as a mouse but as cool as a lion.
 Clear the way, &c.

While the Keystone boys rushed nobly forth,
To stand by the state—the prop of the North,
Brave Mac gave the foe a blow in the rear,
Which made their Stonewall quake with fear.
 Singing, clear the way, brave Mac is advancing

Upon the heights of old "South Mountain,'
He made their blood flow like a fountain,
Till old secession's wings were clipt,
And Lee confessed himself "well whipped,'
 Singing, clear the way, &c.

With HOOKER, FRANKLIN, and brave BURNSIDE,
He nobly turned the battle's tide,
And drove them across the blue Potomac,
With battered heads and an empty stomach.
 Singing, clear the way, &c.

The gallant RENO nobly fell,
But millions of brave hearts toll his knell,
And honor and glory crown the pall,
Of those who for the Union fall.
 Singing, clear the way, &c

At Sharpsburg next, our gallant "Mac,"
Quickly followed suit and beat them back,
He seized their Longstreet by the knob,
And shelled the corn from Howell Cobb.
 Singing, clear the way, &c.

At Harpers Ferry next he sold them,
And made the place "too hot to hold them."
He bound by death or victory,
To set old Maryland safe and free.
 Singing, clear the way, &c.

Three cheers for our glorious hero, "Mac,"
And the gallant army at his back,
He's bound to march to victory forth,
Till the Union Flag floats South and North.
 Singing, clear the way, &c.

More Civil War books like this one in the Nation Divided series, as well as other histories, are available at the Blue Mustang Press website:

www.BlueMustangPress.com

or at your local or online bookseller.

9 781935 199137